SOLAR PROMINENCES

GEOPHYSICS AND ASTROPHYSICS MONOGRAPHS

AN INTERNATIONAL SERIES OF FUNDAMENTAL TEXTBOOKS

Editor

B. M. McCORMAC, *Lockheed Palo Alto Research Laboratory, Palo Alto, Calif., U.S.A.*

Editorial Board

VOLUME 12

SOLAR PROMINENCES

by

EINAR TANDBERG-HANSSEN

*High Altitude Observatory, National Center for Atmospheric Research,
Boulder, Colo., U.S.A.*

D. REIDEL PUBLISHING COMPANY

DORDRECHT-HOLLAND / BOSTON-U.S.A.

Library of Congress Catalog Card Number 73–88593

Cloth Edition: ISBN 90 277 0399 X
Paperback edition: ISBN 90 277 0400 7

Published by D. Reidel Publishing Company,
P.O. Box 17, Dordrecht, Holland

Sold and distributed in the U.S.A., Canada and Mexico
by D. Reidel Publishing Company, Inc.
306 Dartmouth Street, Boston,
Mass. 02116, U.S.A.

To Else and Karin

TABLE OF CONTENTS

PREFACE

O beauté sans seconde
Seule semblable à toi
SOLEIL pour tout le monde . . .
JEAN-FRANÇOIS SARASIN (1615–1654)

The last decade has seen the publication of monographs covering most areas of solar activity: flares (Smith and Smith, 1963), sunspots (Bray and Loughhead, 1964) and the corona (Billings, 1966). Consequently, of all the major manifestations of solar activity only prominences are without a comprehensive and unified treatment in the current literature.

The present book is written in an attempt to remedy this situation, and to furnish an account of some of the most spectacular and most beautiful aspects of solar activity. Our ultimate aim is an understanding of the physical processes involved. I hope that this book may provide if only a small step toward this goal.

After an historical introduction and some general definitions Chapter I proceeds with an account of several classification schemes for prominences. Most of the observational material is presented in Chapter II and forms the basis on which different models of prominences are built in Chapter III. Chapters IV and V give most of the physics of prominences, treating, as they do, the formation and stability of these objects. The interaction of prominences with other manifestations of solar activity is the subject of Chapter VI, and the final Chapter VII considers prominences in the larger context: as an integral part of the corona.

The book is intended for the advanced student in astrophysics who may want to specialize in the fascinating field of solar activity. In addition, I hope that some of my colleagues may find parts of the book helpful in their teaching and research.

It is a pleasure to acknowledge the valuable help of J. Goff in preparing photographs for the figures, and of the Graphic Arts Department of NCAR for all the drawings. Special thanks are due to Mrs R. Fulk for her careful typing of the manuscript.

I have benefited greatly from discussions with many of my colleagues, to whom I give my grateful thanks: R. G. Athay, A. Bruzek, R. T. Hansen, E. Hildner, T. Hirayama, C. L. Hyder, Y. Nakagawa, R. Noyes, G. Pneuman, A. Poland, F. Q. Orrall, E. Priest, and D. Smith.

Finally, I would like to express my appreciation to the astronomers who provided me with some of the illustrations: Mrs S. F. Martin, R. Dunn, and G. Pneuman.

Boulder, Colorado, April 1973

LIST OF SYMBOLS AND PHYSICAL CONSTANTS

a ratio between collisional halfwidth and Doppler halfwidth

A Einstein coefficient; area; amplitude

b parameter expressing deviation from local thermodynamic equilibrium

B Einstein coefficient; magnetic field; Planck function

c velocity of light; specific heat

C collisional rate coefficient

d distance

D distance

e base of natural logarithm; electronic charge

E electric field; energy

f oscillator strength; distribution function

F flux

g gravitational acceleration; statistical weight; Gaunt factor; Landé's g factor

G constant of gravitation; energy gain function

h Planck's constant; height

H scale height; radiative flux

i imaginary unit ($\sqrt{-1}$)

I Stokes parameter; specific intensity

j current density; emission coefficient

J mean intensity; Bessel function; action variable

k Boltzmann's constant; absorption coefficient; wave number

K degree absolute; thermal conductivity; restoring force

l characteristic length

L characteristic length; orbital angular momentum quantum number; energy loss function

m mass

M mass; magnetic quantum number

n principal quantum number; number density of particles; refraction index; vector normal

N number of particles in column of cross section 1 cm^2

p pressure; momentum

P rate coefficient ($P=R+C$); probability; power; period $=2\pi/\omega$; degree of polarization

q heat flow

Q Stokes parameter; energy

r cylindrical polar coordinate; distance (radial); ratio of continuum absorption coefficient to line absorption coefficient

R gas constant; radiative rate coefficient; Reynolds' number
s entropy; coordinate along field line
S spin quantum number; source function; Poynting vector
t time
T temperature; stress tensor
u velocity
U partition function; heat energy; Stokes parameter
v flow velocity (macroscopic)
V gross velocity; wave velocity; volume; Stokes parameter
w particle velocity (microscopic)
W energy; geometrical dilution factor
x cartesian coordinate
X prominence thickness
y cartesian coordinate
z cartesian and cylindrical polar coordinate
Z atomic number

α filamentary structure ratio; coefficient of volume expansion; absorption coefficient per atom; angle; reciprocal pitch
γ ratio of specific heats; damping constant; angle; ratio between gas pressure and magnetic pressure
δ delta function; angle
$\Delta\lambda$ halfwidth of spectral line
ε emissivity; dielectric constant; energy density
ε' complex dielectric constant ($\varepsilon' = \varepsilon - i(4\pi\sigma/\omega)$)
η coefficient of viscosity
θ angle
κ continuum quantum number
λ wavelength; mean free path
λ_B Larmor radius
μ magnetic moment; mean molecular weight
ν frequency
ξ microturbulent velocity
ρ density; net radiative bracket
σ Stefan's constant; electric conductivity
τ optical depth; diffusion time
ϕ cylindrical polar coordinate; latitude; phase; absorption profile; scalar potential; angle variable
Φ magnetic flux; gravitational potential
χ electric potential
ω circular frequency; angular velocity

Physical Constants

Boltzmann's constant $k = 1.38 \times 10^{-16} \text{ erg K}^{-1}$
Electron rest mass $m_e = 9.1 \times 10^{-28} \text{ g}$

Elementary charge	e	$= 4{\cdot}80 \times 10^{-10}$ esu
Gravitational constant	G	$= 6{\cdot}67 \times 10^{-8}$ dyne cm^2 g^{-2}
Planck's constant	h	$= 6{\cdot}626 \times 10^{-27}$ erg
Proton rest mass	m_p	$= 1{\cdot}67 \times 10^{-24}$ g
Stefan-Boltzmann constant	σ	$= 5{\cdot}67 \times 10^{-5}$ erg cm^{-2} s^{-1} K^{-4}
Velocity of light	c	$= 2{\cdot}998 \times 10^{10}$ cm s^{-1}

ABBREVIATIONS FOR JOURNALS

Astron. J.	Astronomical Journal
Astron. Zh.	Astronomischeskii Zhurnal
Ann. Astrophys.	Annales d'Astrophysique
Astrophys. J.	Astrophysical Journal
Astrophys. Norv.	Astrophysica Norvegica
Bull. Astron. Inst. Czech.	Bulletin of the Astronomical Institute of Czechoslovakia
Bull. Astron. Inst. Neth.	Bulletin of the Astronomical Society of the Netherlands
Compt. Rend. Acad. Sci.	Comptes Rendus (Paris)
J. Geophys. Res.	Journal of Geophysical Research
Monthly Notices Roy. Astron. Soc.	Monthly Notices of the Royal Astronomical Society
Publ. Astron. Soc. Japan	Publications of the Astronomical Society of Japan
Publ. Astron. Soc. Pacific	Publications of the Astronomical Society of the Pacific
Proc. Astron. Soc. Australia	Proceedings of the Astronomical Society of Australia
Solar Phys.	Solar Physics
Z. Astrophys.	Zeitschrift für Astrophysik

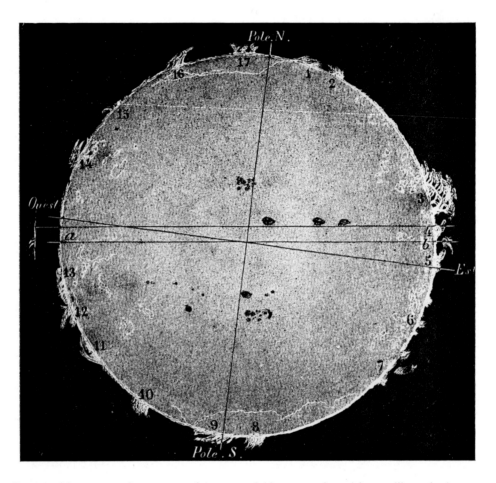

Les protubérances se présentent sous des aspects si bizarres et si capricieux qu'il est absolument impossible de les décrire avec quelque exactitude.

SECCHI, *Le Soleil,* 1877.

CHAPTER I

INTRODUCTION

1.1. Historical Background

In 1239 Muratori (see Secchi, 1875) observed the corona during a total eclipse and reported 'a burning hole' in it. This burning hole in all probability was a prominence, and Muratori's report is one of the earliest we have of this sign of solar activity. Medieval Russian chronicles (see Vyssotsky, 1949) also mention prominences, but the first semi-scientific description of them came after the eclipse of May 2, 1733. During this event Vassenius (1733) observed three or four prominences from Gothenburg, Sweden. He called them 'red flames', and believed them to be clouds in the lunar atmosphere. Celsius (1735) edited a report of the Swedish observations which shows that other observers agreed with Vassenius' description (see also Grant, 1852). Ulloa (1779) observed what probably was an active prominence during the eclipse of 1778, and attributed it to a hole in the Moon.

These early observations of prominences were subsequently forgotten, and Bailey, Airy, Struve and Schidlofscky, Arago and others were all taken by surprise when they rediscovered the phenomenon during the eclipse of July 8, 1842 in France and Italy. They were so baffled and amazed that hardly any reliable account is available of what they saw. Hence, their descriptions of the shape of prominences were so vague that they could not prevent later observers from believing that prominences were mountains on the Sun (see Grant, 1852). By contrast, the next eclipse observations made in Norway and Sweden in 1851 led to uniform interpretations and conclusions, and the theory that prominences were mountains could be ruled out (see Secchi, 1875; Young, 1896).

Two great steps forward were taken in 1860 at the eclipse in Spain when Secchi (1875) introduced photography into the observations, and in 1868 at the August 18 eclipse in India and Malacca when spectrographic methods were employed for the first time (Secchi, 1868; de la Rue, 1868). Spectra of prominences were found to consist of bright lines, and from then on prominences were considered to be glowing masses of gas. One of the observed lines, at 5876 Å (the D_3 line), was not known to be emitted by any terrestrial atom, and it was ascribed to a specific solar element, called He, after Helios, the Greek sun god.

An interesting development took place immediately following the eclipse of August 18, 1868. Janssen (1868) realized that many of the emission lines that he had just observed were so bright that they should be visible without the help of an eclipse. To test this, the next day he aimed the spectrograph slit outside the limb of the sun's image and observed prominence emission lines in full daylight. The same kind of observation was reported independently and simultaneously by Lockyer (1868a), who with Huggins had tried to accomplish this for some time. Since then regular observations of prominences have been carried out without eclipses, but for more than 60 yr the special conditions made eclipse observations of prominences superior, and they continued to be of great importance.

At the time of the eclipse of 1869 in North America, Huggins (1869) realized that by opening the spectograph slit one could obtain a series of monochromatic images of prominences, corresponding to the emission lines observed with a normal slit. By using this method the complex forms of prominences could be better studied.

In an important investigation, Schwarzschild (1906) published the results of the first systematic photometric measurements of spectral lines in prominences, using the data from the August 30, 1905 eclipse. Since that time, observations during many later eclipses, as well as limb observations performed outside eclipses, have gradually added to our knowledge about the shapes, dimensions, distribution on the sun, spectrum, movements, and changes of prominences (see for instance Pettit (1919, 1932), Davidson and Stratton (1927), Pannekoek and Doorn (1930), Grotrian (1931), Mitchell (1935), and McMath and Pettit (1937, 1938)). In the 1890's spectrohelio-graphs became available and prominences could be studied on the disk as absorption features, generally referred to as filaments (see Deslandres, 1910; Hale and Ellerman, 1903).

Finally, with Lyot's invention of the coronagraph, it became possible to observe limb prominences at any time nearly as thoroughly as during eclipses (Lyot, 1936, 1937, 1939).

We could also quote from the vast literature after 1940, but these papers will be referred to as we discuss the different detailed aspects of prominences in the following sections.

It may be worthwhile to conclude this brief history of prominence observations by a quote from the truly remarkable book *Le Soleil* by Secchi (1875). He began his chapter on prominences with the following statement: "The phenomenon of prominences is now so well known by everybody that it may seem unnecessary to retrace the history of their discovery." It is sobering to note that more than ninety years later 'the well-known phenomenon' still poses some of the most puzzling questions in solar activity.

1.2. Definitions

The term prominence is used to describe a variety of objects, ranging from relatively stable structures with lifetimes of many months, to transient phenomena that last but hours, or less. It is not easy to give a short, concise definition of prominences that encompasses the necessary and sufficient criteria, the difficulty being to distinguish them from some flares. When seen projected against the solar disk, nearly all prominences show up in absorption, looking dark against the bright disk, while flares show in emission, being brighter than the normal, quiescent disk. However, there are active, short-lived prominences that show up in emission, at least during certain stages of their development. When observed above the solar limb, prominences invariably show up bright against the dark sky background, as do flares. An often used definition of a prominence is an object in the chromosphere or corona that is denser and cooler than its surroundings. However, parts of many flares fit the same definition. Being aware of this difficulty, we shall think of prominences as cool, dense objects in the hotter corona. Exceptions to this crude definition will be treated as we explore the physical conditions reigning in prominences.

We shall refer to the long-lived (days to months), only slowly changing prominences seen away from active regions with sunspots as quiescent prominences (see Figure I.1 for an example). When these are seen in absorption against the disk, they are often called filaments. In and around active regions different kinds of more short-lived, rapidly changing prominences occur (see Figure I.2). We refer to these as active region prominences, or simply active prominences. There are intermediate types, prominences that do not fall easily into either the definitely quiescent or the definitely active class. We shall return to these in Section 1.3.

So far as the different types of prominences are concerned, there is no universally adopted definition, or even description, for many of them. It is easy to run into one of two difficulties: either too coarse a classification is given, so that important variations between different types of prominences cannot be utilized, or one devises so fine a classification that the physics is drowned in insignificant observational details. We shall here try to describe, and thereby to some extent define, what we shall mean when we talk about the more common and important types of prominences. First, however, let us define some commonly used terms from the vocabulary of solar activity.

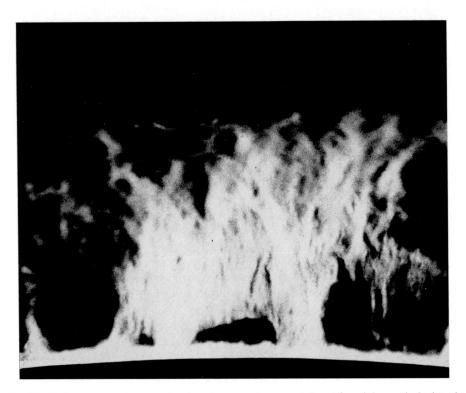

Fig. I.1. Quiescent prominence, showing fine structure consisting of mainly vertical threads. (Courtesy Dr R. Dunn, Sacramento Peak Observatory.)

Fig. I.2. Active prominences with material streaming (often nearly horizontally) toward distant centers of activity. Climax station of the HAO.

Photospheric faculae are areas seen on white-light photographs of the Sun, that are brighter than the surrounding photosphere.

Chromospheric faculae are areas on photographs taken in the nearly monochromatic light of a strong spectral line, like Hα or the Ca II, K-line, that are brighter than the surrounding chromosphere. The French notation is *plage faculaire*, and the term *plage* will be adopted as an alternative name for a chromospheric facula. There is a continuous transition as one goes out in the atmosphere from the deep-lying photospheric faculae to the plages. The two types resemble each other in occurrence, shape and their relation to sunspots. Sunspots nearly always occur in pre-existing faculae.

Active regions are areas of the Sun's atmosphere where excess magnetic flux is found. The magnetic field causes a local heating of the atmosphere, and this is observed as a plage. Above the optical plage other signs of activity are often observed. The corona is hotter and denser (coronal enhancements and condensations), leading to excess emission of forbidden coronal lines, radio waves, X-rays and particles.

Of the more commonly seen prominence types we mention:

Eruptive prominences (also called ascending prominences) are ordinary quiescent prominences that for some reason become unstable, erupt (ascend) and disappear. The French term for this phenomenon is *disparition brusque,* and one often refers to the disparition brusque phase of a quiescent prominence. Generally, the prominence reforms in the same place. According to L. d'Azambuja and M. d'Azambuja (1948), the disparition brusque phase seems to be a normal stage in the development of most quiescent prominences.

Fibrils are long thin dark threads visible in Hα on the disk at the edge of plages and near the edge of sunspot penumbra (Lippincott, 1955).

Arch filaments are similar to fibrils, but larger and longer (20000 to 30000 km). They are found early in the life of most active regions, characteristically in the inter-spot region of developing bi-polar spot groups (Bruzek, 1967, 1969a).

Active region filaments (also called active sunspot prominences) are seen as very dark features on the disk inside active regions. They look superficially somewhat like small ordinary, quiescent filaments. They have lifetimes of days and terminate in sunspots at one or both ends.

Caps are seen above the limb as bright, low-lying prominences near active regions. Their lifetimes range from hours to days. Surges frequently are ejected from the edges of caps (Pettit, 1943). They may be the limb manifestation of fibrils and arch filaments (Malville, 1968; Harvey, 1969).

Knots are seen above the limb as shortlived (15 min) very bright features, having heights less than 20000 km. They lie above sunspots.

Surges are prominences that seem to be shot out of active regions as long straight or curved columns, and return along the same trajectory Figure I.3. They may reach to great heights (several hundreds of thousands of km) and their velocities may exceed several hundred km s^{-1}. Some active regions produce nearly identical surges during part of their lives (homologous surges).

Sprays are shot out from flare regions at velocities often exceeding the velocity of escape. The ejection is so violent that the matter is not contained, as in surges, but flies out in fragments.

Coronal clouds are irregular objects suspended in the corona with matter streaming out of them into nearby active regions along curved trajectories. The coronal clouds last for a day or more at heights of several tens of thousands of km.

Loops. As the name suggests these prominences have a loop structure and their feet are anchored in or near sunspots. They occur as the result of a major flare. Material is generally seen to stream down the two legs of the loop. A loop system is the manifestation of the highest degree of activity observed optically in the solar atmosphere. At the tops of such loops the corona is very hot and condensed into a coronal condensation.

Coronal rain is closely related to loops, but the complete loop structure is absent, giving the phenomenon its descriptive name.

1.3. Morphological Classifications

Prominences can be classified in several ways. From the preceding sections it follows that prominences may take on very different forms, their lifetimes may range from minutes to many months, and the degree of dynamic activity varies greatly from one prominence to another. It would seem natural to classify prominences using some of these characteristics, and, indeed, Secchi (1875) already divided them into quiescent and active (he called the latter eruptive) prominences. Secchi further sub-divided the

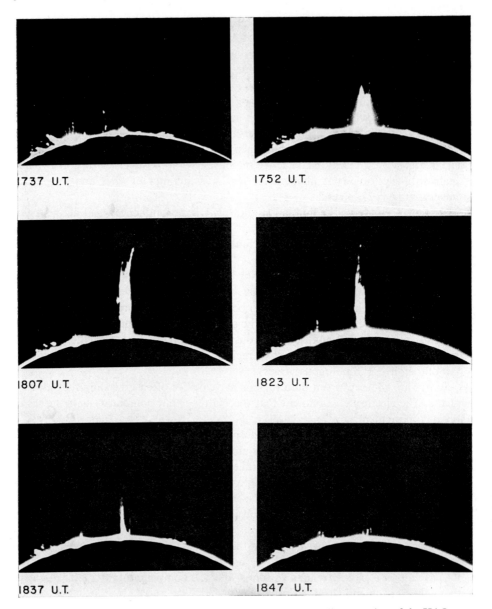

Fig. I.3. Development of surge prominence, June 12, 1946. Climax station of the HAO.

quiescent and active prominences into subclasses (clouds, filaments, stems, plumes, horns, cyclones, flames, jets, sheafs, spikes – see also Young, 1896). It is difficult to maintain the distinction between these subgroups, but the main two-class arrangement is of lasting value, and we still use it as the basic classification.

There are also significant differences in the spectra of quiescent and active prominences. How this fact can be used to furnish classification criteria will be treated in the next chapter.

In the last few years magnetic fields have been observed in prominences, and it is conceivable that this may lead to another useful classification scheme. However, this branch of prominence research is still in its early infancy, and only a rudimentary classification can presently be deduced from it (see Chapter II).

We shall now treat the more pragmatic, morphological classifications based on the topography and motions of prominences.

1.3.1. PETTIT'S CLASSIFICATON

For many years the most widely known and used classification was that due to Pettit (1925, 1932, 1936, 1943, 1950). He divided prominences into five classes, as shown in Table I.1. Classes 1 and 2 are closely related according to Pettit; a given prominence

TABLE I.1

Pettit's classification

Class	Name	Description
1	Active	Material seems to be streaming into nearby active center (like sunspots).
2	Eruptive	The whole prominence ascends with uniform velocity (of several hundred km s^{-1} often). The velocity may at times suddenly increase.
3	Sunspot	These are found near sunspots and take the shape of 'water in a fountain' or loops.
4	Tornado	A vertical spiral structure gives these prominences the appearance of a closely wound rope or whirling column.
5	Quiescent	Large prominence masses which show only minor changes over periods of hours or days.

may for instance pass from the active to the eruptive state. Class 3 contains some of the most dynamically active objects. The tornado prominences of class 4 seem to be very rare objects, only occasionally discussed in the literature (Harvey, 1969). Pettit's class 5 are the classical quiescent prominences that take the form of enormous sheets standing vertically in the solar atmosphere. Characteristically, the sheets touch the solar surface only at certain fairly regularly spaced intervals, which we may call the feet of the prominence (see Figure I.1). The dimensions of these, the largest of all prominences, are about 200000 km in length, 50000 km in height but only 5000 to 10000 km in thickness. No prominence is really quiescent, but this type shows slower and less pronounced changes than any other class. However, we shall see later (Chapter VI) that they may go through some active phases in their lifetimes (prominence activations).

1.3.2. NEWTON'S CLASSIFICATION

It was realized early both by Hale (Hale and Ellerman, 1903) and by Deslandres (1910) that the dark filaments seen in absorption on the disk in their spectrohelio-grams are nothing more than prominences seen against the bright photosphere. However, further progress along these lines was hampered by two drawbacks of the spectroheliograph, viz. (1) it takes considerable time to obtain a spectroheliogram and get ready for the next, and (2) for prominences with motions in the line-of-sight the Doppler effect will throw the spectral line outside the second slit of the instrument, and the observation will be lost. When Hale (1929, 1930, 1931a, 1931b) constructed the spectrohelioscope, both these drawbacks were remedied. With the spectrohelioscope the observations are made visually and the line is kept on the second slit by a line-shifter whose position gives directly the radial velocity. Newton (1934, 1935) used this technique to study filaments and derived the following classification (see Table I.2).

TABLE I.2

Newton's classification

Class	Description
I	Prominences that avoid the neighborhood of sunspots (but not the whole sunspot zone). Long well-defined filaments lasting several days.
II	Prominences that are associated with sunspots or with the plage areas. Generally smaller than objects of class I. Lifetimes of the order of minutes or hours.

(a)	(b)
Prominences that show large radial velocities and occur after the appearance of a localized emission (a so-called bright flocculus).	Prominences originally of class I that become activated by the sudden appearance of an emission object. The emission remains more or less stationary, but it gives the filament a large radial velocity.

The objects of class I are identical to Pettit's quiescent prominences, while Pettit's classes 1, 2, 3 and 4 are all comprised in Newton's class II – active prominences. We note that Newton's class IIb describes an interesting flare-prominence relationship, which we shall study in more detail in Chapter VI.

1.3.3. The Menzel-Evans classification

Menzel and Evans (1953) introduced an important two-dimensional classification based on (1) whether the prominence seems to originate from above, i.e. in the corona (A), or from below (B), and (2) whether the prominence is associated with sunspots (S) or not (N). The further subdivision is pictorial, based on the shapes of the objects (see Table I.3). Each type of prominence is further designated by a lower case letter.

TABLE I.3

The Menzel-Evans classification

Relation to Sunspots	Place of origin	
	From above, A	From below, B
Associated with sunspots, S	Rain, a Funnels, b Loops, l	Surges, s Puffs, p
Not associated with sunspots, N	Coronal rain, a Tree trunks, b Trees, c Hedgerows, d Suspended clouds, f Mounds, m	Spicules, s

For example surges are referred to by the symbols BSs, and hedgerows (quiescent prominences) by ANd. The distinction between A and B type prominences is fundamental for the discussion of prominence formation (Chapter IV).

1.3.4. SEVERNY'S CLASSIFICATION

Severny (1950, 1959) and Severny and Khokhlova (1953) argued that to arrive at a classification of any physical significance, one must characterize prominences mainly by their motions. In this scheme quiescent prominences are not given separate attention, they belong to class III, irregular prominences, while class II corresponds to sunspot prominences in the Menzel-Evans classification (see Table I.4). Again, it is apparent that the word quiescent is to be taken in a relative sense only.

TABLE I.4

Severny's classification

Class	Name	Description
I	Eruptive	Quiescent prominences becoming eruptive. Rare, 5 to 10% of all cases. Outward motions with velocities of several hundred km s^{-1} at times exceeding the velocity of escape.
II	Electromagnetic	Electromagnetic prominences. The knots or condensations making up the prominences exhibit motions along definite curved trajectories. Velocities range from several tens to a few hundred km s^{-1}. About 50% of all prominences belong here.
III	Irregular	Prominences with irregular, random, motions of individual knots.

1.3.5. DE JAGER'S CLASSIFICATION

De Jager (1959), realizing, as did Severny and Khokhlova, the importance of motions, divides prominences into two main groups: quiescent and moving prominences. In several ways the word 'moving' may be a better attribute than 'active'; however, only in a relative sense are the quiescent prominences non-moving. The finer subdivisions follow fairly closely Pettit's classification (see Table I.5). De Jager distinguishes

TABLE I.5

De Jager's classification

Class	Name
I	Quiescent prominences (a) Normal (low to medium latitudes) (b) Polar (high latitudes)
II	Moving prominences (a) Active (b) Eruptive (c) Spot (d) Surges (e) Spicules

between normal quiescent prominences and quiescent prominences found in high latitudes. Ultimately these latter prominences, which drift polewards, form the so-called polar crowns. Furthermore, de Jager classifies explicitly the important prominence types surges and spicules, as did Menzel and Evans.

1.3.6. ZIRIN'S CLASSIFICATION

Zirin (1966) divides prominences into two sharply defined classes depending on their relation to solar activity – in particular whether they are related to flares or not. Those that are related to flares, class I, exhibit violent motions and are short-lived (see Table I.6). Class II are the long-lived quiescent prominences, subdivided in the same manner

TABLE I.6

Zirin's classification

Class	Description	Object
I	Short lived, associated with flares and active sunspots	1. Sprays, explosions, puffs 2. Surges 3. Loops, coronal rain
II	Long lived, quiescent	1. Polar cap filaments 2. Sunspot zone filaments
III	Intermediate	1. Ascending prominences 2. Sunspot filaments

as de Jager's class I. In addition, Zirin defines an intermediate class III comprising quiescent prominences during their disparition brusque phase. Also, this class explicitly classifies the important dark filaments seen in active regions. Their relationship to different aspects of solar activity will be treated in Chapter VI.

CHAPTER II

OBSERVATIONAL DATA

We have already discussed some preliminary observational data in Chapter I when describing the morphological classifications of prominences. However, these observations pertained mainly to the easily observed shapes and overall motions of the prominences. In this chapter we shall in more detail consider these and other data and see how they can lead to physical models of different kinds of prominences.

2.1. Spectra

2.1.1. INTRODUCTORY REMARKS

A wealth of information is contained in the visible part of spectra of prominences as observed above the limb, and since most of our information on the physics of prominences stems from such spectral observations, we shall spend some time on this subject. For radio, EUV and X-ray observations, see Section 2.6. If the prominence plasma is dense enough, i.e., when the prominence looks very bright, so that there is a sufficient number of atoms in the line of sight, N cm^{-2}, the number of spectral lines from H, He, and metals are impressive indeed. Secchi (1875) already was aware of this, and since his time spectra of prominences have been studied by a series of authors, including more recently Grotrian (1931), Richardson (1950), Conway (1952), Ananthakrishnan and Madhavan Nayar (1953), ten Bruggencate (1953), Severny (1954), Haupt (1955), Ivanov-Kholodny (1955, 1958), Zuikov (1955), ten Bruggencate and Elste (1958), Sobolev (1958), Jefferies and Orrall (1958, 1961a, 1961b, 1961c, 1962), Tandberg-Hanssen and Zirin (1959), Zirin and Tandberg-Hanssen (1960), Zirin (1959), Tandberg-Hanssen (1959, 1963, 1964), Bratiichuk (1961), Rigutti and Russo (1961), Shin-Huei (1961), Moroshenko (1963), Hirayama (1963, 1964, 1971a), Yakovkin and Zel'dina (1963, 1964a, 1964b), and Stellmacher (1969).

As mentioned in Section 1.3 there are significant differences in the spectra from quiescent and from active prominences. The omnipresent H lines are difficult to use to distinguish different types of prominences, since they often are overexposed when one tries to bring out the weaker metal lines. Besides, problems of self-absorption render the analysis uncertain. Both He and metal lines are found in all types of prominence spectra, albeit with greatly varying intensities. From a comparison of relative intensities and halfwidths one can devise useful classification schemes. Figure II.1 shows the difference between part of a spectrum from a quiescent prominence (Sept. 22, 1962) and from an active prominence (Febr. 19, 1962). Notice that in the spectrum of the active prominence the intensity of the line of He II (at 4686 Å) is comparable to the intensity of the He I line (at 4713 Å). In all quiescent prominences lines of He II are quite faint. On the other hand, the many metal lines, in particular

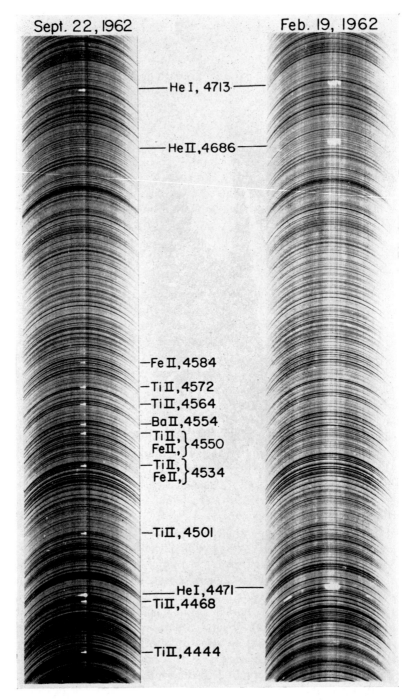

Fig. II.1. Spectra of quiescent (Sept. 22, 1962) and active (Febr. 19, 1962) prominences.

from Fe II and Ti II, which are easily seen in the spectra of quiescent prominences, are faint in spectra of active prominences. However, they are generally present also in the active objects, and the reason for their weakness is simply that the number of atoms in the line-of-sight in the right excitation and ionization condition is too small to result in easily observable lines. At times the active prominences are bright enough, i.e., the prominences are of sufficient extent or are dense enough, so that the metal lines in their spectra are conspicuous. These are probably the type of prominences that Secchi (1875) and Ananthakrishnan and Madhavan Nayar (1953) call metallic, a nomenclature that we shall abandon, since it may imply incorrect conclusions regarding the physics of these objects.

2.1.2. BASIC NOTATIONS – OPTICAL SPECTRA

The intensity of an emission line is given by

$$I_v = \int S_v \, e^{-\tau_v} \, d\tau_v \,, \tag{II.1}$$

where S is the source function at frequency v, τ_v the optical depth, $\tau_v = \alpha_v N$, and α_v is the absorption coefficient per atom. N is the number of atoms in the line of sight (cm^{-2}). At the line center ($v = v_0$, or $\lambda = \lambda_0$) the absorption coefficient is

$$\alpha_0 = \frac{\pi e^2}{mc} \frac{f}{\sqrt{\pi} \, \Delta v_D} \,, \tag{II.2}$$

where Δv_D is the Doppler halfwidth. The analysis of Equation (II.1) involves the fundamental question of the nature of the source function. In most work on prominences S_v has been assumed frequency independent and constant with optical depth in the region where the line is formed. Only under this simplifying assumption does Equation (II.1) yield the line profile

$$\frac{I_v}{I_0} = \frac{1 - e^{-\tau_v}}{1 - e^{-\tau_0}} \,, \tag{II.3}$$

where an index zero indicates the value at the line center.

The shape of the absorption coefficient depends on the dominant line-broadening mechanism. We find that the most important source of line broadening in prominences is the Doppler effect, and the absorption coefficient is given by

$$\alpha_v = \alpha_0 \exp \left(- \frac{\Delta \lambda}{\Delta \lambda_D} \right)^2 \,, \tag{II.4}$$

where $\Delta \lambda = \lambda - \lambda_0$, the Doppler halfwidth $\Delta \lambda_D = (\lambda/c) \xi_0$ and ξ_0 is the line-broadening velocity. If the prominence is optically thin in the observed line, the shape of the line will be Gaussian, i.e., the line profile will be Doppler broadened

$$\frac{I_v}{I_0} \propto \exp \left(- \frac{\Delta \lambda}{\Delta \lambda_D} \right)^2 \,. \tag{II.5}$$

If the broadening is entirely due to the thermal broadening of the atoms (or ions), the Doppler width of the line is given by

$$\Delta\lambda_{\mathrm{D}} = \frac{\lambda}{c}\sqrt{\frac{2kT}{m}} , \tag{II.6}$$

where m is the mass of the atom and T the kinetic temperature of the plasma. In general, the atoms possess a non-thermal velocity component ξ_t (the astronomical 'micro-turbulence'), whose influence on the line profile mimics that of the thermal velocity, giving rise to a total Doppler width

$$\Delta\lambda_{\mathrm{D}} = \frac{\lambda}{c}\sqrt{\frac{2kT}{m}+\xi_t^2} . \tag{II.7}$$

In principle, we can separate the thermal and non-thermal velocity components and solve for the kinetic temperature and the turbulent velocity ξ_t, provided we can determine the widths of two lines from different elements, i.e., different masses, $m_i = \mu \cdot m_{\mathrm{H}}$, $i = 1, 2$, where μ is the atomic weight and m_{H} the mass of a H atom. This procedure is, of course, only permissible if the two lines are excited in the same volume element of the prominence plasma. From Equation (II.7) we then find

$$T = 1 \cdot 95 \times 10^{12} \left(\frac{1}{\mu_1} - \frac{1}{\mu_2}\right)^{-1} \left[\left(\frac{\Delta\lambda_{\mathrm{D}_1}}{\lambda_1}\right) - \left(\frac{\Delta\lambda_{\mathrm{D}_2}}{\lambda_2}\right)^2\right] , \tag{II.8}$$

and

$$\xi_t = 3 \cdot 24 \times 10^{20} \left(\frac{\Delta\lambda_0}{\lambda}\right)^2 - 1 \cdot 64 \times 10^8 \frac{1}{\mu} . \tag{II.9}$$

It should be pointed out that indiscriminate use of Equations (II.8) and (II.9), when there was no obvious reason to expect the two lines to come from the same region of the prominence plasma, has at times led to unintelligible results.

Even though the Doppler effect in general is the most important line-broadening mechanism in prominences, we find that for many H lines collisional broadening may at times contribute significantly to the line shape. The effects of collisions are complicated, and two limiting cases, collisional damping and statistical broadening, are generally considered. For details on these approximations, which also are called the impact and the quasistatic approximations, see for instance Jefferies (1968). The collisional damping, which mainly pertains to the core of the line, can formally be treated like the natural broadening (radiation damping) of the line, and the absorption coefficient per atom is

$$\alpha_\nu = \alpha_0 \frac{\gamma}{4\pi^2(\nu-\nu_0)^2+(\gamma/2)^2} , \tag{II.10}$$

where the damping constant $\gamma = \gamma_{\mathrm{rad}} + \gamma_{\mathrm{coll}}$ is the reciprocal of the mean lifetime of the upper level of the transition giving rise to the line in question. The resulting line shape is described by a symmetrical dispersion profile, and in all practical cases the collisional broadening dominates the radiation damping. We have a Lorentz broadened line

$$\frac{I_\lambda}{I_0} \propto \frac{\Delta\lambda_{\mathrm{c}}}{(\lambda-\lambda_0)^2+(\Delta\lambda_{\mathrm{c}})^2} , \tag{II.11}$$

where the damping constant $\Delta\lambda_{\mathrm{c}}$ is a collisional halfwidth.

When Doppler effect due to thermal and turbulent motions as well as natural and collisional damping all are present, the line-absorption coefficient is given in terms of the Hjerting function $H(a,v)$ (Hjerting, 1938), and we have a Voigt broadened line,

$$\alpha_v = \alpha_0 H(a, v) = \alpha_0 \frac{a}{\pi} \int_{-\infty}^{\infty} \frac{e^{-y^2} \, dy}{a^2 + (v-y)^2}, \tag{II.12}$$

where $a \equiv \Delta\lambda_c/\Delta\lambda_D$ and $v \equiv \Delta\lambda/\Delta\lambda_D$. Since a is always small, the $H(a, v)$ function can be expanded in a Taylor series

$$\frac{\alpha_v}{\alpha_0} = H_0(v) + a H_1(v) + a^2 H_2(v) + \cdots \tag{II.13}$$

where

$$H_0(v) = e^{-v^2}$$

$$H_1(v) = -\frac{2}{\sqrt{\pi}} \left(1 - 2v \, e^{-v^2} \int_0^v e^{t^2} \, dt \right)$$

$$H_2(v) = (1 - 2v^2) \, e^{-v^2} \quad \text{etc.}$$

and have been tabulated by Harris (1948).

Solar spectral lines are often analyzed in terms of the Voigt profile. We see from Equation (II.12) that when v is small, i.e., near the line center, $H(a, v) \propto e^{-v^2} = \exp(-\Delta\lambda/\Delta\lambda_D)^2$, and we retrieve a Gaussian shape. On the other hand, in the wings $H(a, v) \propto v^{-2} \propto 1/\Delta\lambda^2$, and the damping, or Lorentz, profile holds. For all practical purposes the absorption coefficient is Gaussian out to about 3 Doppler widths, and we shall often use this fact in analyzing prominence lines.

The effect of statistical broadening, which pertains to the wings of the line, is treated in an average way by the theory of the Stark effect. The interaction of the ions and electrons with the emitting atom or ion is expressed in terms of a mean interionic electric field, F_0, and in the wing of the line the absorption coefficient is approximated by

$$\alpha_v \propto \frac{F_0^{3/2}}{(\Delta\lambda)^{5/2}}. \tag{II.14}$$

In Holtsmark's approximation the field F_0 is a simple function of the ion or electron pressure, and Equation (II.14) can be written

$$\alpha_v \propto \frac{n_e}{(\Delta\lambda)^{5/2}}. \tag{II.15}$$

Hence, at sufficient distances from the line center the Stark-broadened profile takes the form

$$\frac{I_\lambda}{I_0} \propto (\Delta\lambda)^{-5/2}. \tag{II.16}$$

When we compare Equation (II.16) with Equation (II.5), we see how a study of the line wings can be used to decide which line-broadening mechanism is at work.

We now combine the effect of thermal and turbulent motions, which dominate the core, with the effect of Stark broadening on the line. In other words, we widen the Stark profile, Equation (II.14) with the Doppler profile, Equation (II.4), and arrive at an expression fror the absorption coefficient of the form

$$\frac{\alpha_v}{\alpha_0} \propto \int_{-\infty}^{\infty} v^{5/2} e^{-v^2} \, dv \, . \tag{II.17}$$

The corresponding profiles have been evaluated by de Jager (1952).

2.1.3. HALFWIDTHS

Analyses of halfwidths of different spectral lines can furnish important information on the physics of the emitting prominence plasma. We shall now consider different types of prominences to see how they can be distinguished by the shape of their spectral lines, which are assumed to be Doppler broadened. In other words, we omit for the time being higher Balmer lines where Stark broadening is of importance.

(a) Quiescent Prominences

Most quiescent prominences exhibit narrow spectral lines with halfwidths corresponding to a broadening velocity ξ_0 of between 5 and 10 km s^{-1}. This holds true fcr H and metal lines as well as for HeI lines (ten Bruggencate and Elste, 1958; Shin-Huei, 1961; Hirayama, 1963, 1971a; Yakovkin and Zel'dina, 1971). Under the assumption that all the lines are emitted from a prominence plasma characterized by a single temperature and turbulent velocity, we find for quiescent prominences the following typical values for the solution set $\{T, \xi_t\}$ of Equations (II.8) and (II.9)

$$T \approx 6000 \text{ K}$$
$$\xi_t \approx 6 \text{ km s}^{-1}. \tag{II.18}$$

However, at times we observe seemingly quiescent prominences whose He lines are too broad relative to the H and the metal lines to give a unique set of temperature and turbulent velocity. Such prominences seem to consist of different regions, viz. (i) a cool region characterized by values of T and ξ_t similar to those of expressions (II.18), whence originate the H and metal line emission, and (ii) a hot region, or He region, whose temperature and/or turbulent velocity is greater than that of the quiescent region. The line-broadening velocity in these more active regions are typically 12 to 15 km s^{-1}. We call such prominences spectroscopically semi-active. They may show no other signs of activity in a dynamical sense of the word.

We are thus led to adopt a *multi-component* model for the prominence plasma, a concept we shall return to shortly.

(b) Active Prominences

The lines in active prominences are markedly broader than in quiescent prominences. Often the lines retain a Gaussian shape, and the broadening velocity ranges from values similar to those in semi-active prominences to values in excess of 50 km s⁻¹. At other times the line profiles are distorted to such a degree that one must conclude that macroscopic mass motions are responsible. Under such conditions little can be deduced concerning the thermodynamic properties of the prominence plasma. In what follows all conclusions are based on spectra whose lines are Gaussian, unless otherwise stated.

The analysis of spectra from active prominences (like loops and surges) leads us to conclude that it is, in general, not possible to find a solution set of T and ξ_t that will satisfy the widths of He lines, as well as H and metal lines. If we combine He and metal lines, Equations (II.9) and (II.10) yield $T \approx 300\,000$–$400\,000$ K. Even more than for semi-active prominences we are forced to consider a multi-component model of the prominences.

(c) Variation of halfwidth with height

The halfwidth $\Delta\lambda_D/\lambda$, or ξ_0, may not be constant over the whole prominence. In quiescent prominences there is a tendency for ξ_0 to be greater in the outer parts, i.e., where the prominence material is closer to the hot coronal plasma (Hirayama, 1964). This effect is more pronounced in active prominences that seem to originate from above. On the other hand, in prominences they are shot up from below (surges) the opposite effect takes place: here the halfwidth decreases with height. Table II.1 shows the variation of halfwidth expressed in terms of the broadening velocity in different kinds of prominences. These data point to the physical reason for the differences between the two kinds of prominences. A change in ξ_0 is to be interpreted as a change in the activity (meaning now temperature and/or motion) of the emitting gas. The data in Table II.1 then show that the temperature and/or motion is less in the upper part

TABLE II.1

Variation of broadening velocity, ξ_0 with height, h, in different kinds of prominences

	Loop			Surge	
	ξ_0 (km s⁻¹)			ξ_0 (km s⁻¹)	
h (km)	He I	He II	h (km)	He I	He II
17000	38	40	8000	57	90
21000	40	52	12500	61	81
25000	50	48	17000	49	—
29000	50	56	21500	49	69
34000	53	43	25000	47	—
38000	—	48			

of surges than near their base. Since matter is being shot up from below, the plasma will gradually relax as it rises and will show less sign of activity the higher we observe it. On the other hand, loops and many quiescent prominences show decidedly more activity near their tops. Loops, in particular, are intimately connected with the hot coronal plasma, from which they seem to derive their energy.

2.1.4. Intensities

(a) *Line Emission – Spectral Classification*

In conjunction with Figure II.1 we pointed out differences between the relative intensities of spectral lines from a quiescent and from an active prominence. While all prominences exhibit strong lines of H (especially early Balmer lines) and of ionized Ca (in particular the K and H lines), the relative intensities of He I and He II and metal lines vary strongly from one type of prominence to another. It would therefore seem natural to try to classify prominences using the relative strength of different spectral lines as criteria. Line strength ratios should, in principle, be measurable for all states of the solar plasma (chromosphere, corona, flares, active prominences, quiescent prominences). The quiet chromosphere might provide the comparison basis, since line-intensity ratios are well known there. The conditions in the chromosphere change with height, from the lower, cooler part below about 1500 km, to the higher chromosphere, and the line-intensity ratios change correspondingly (Thomas and Athay, 1961). For example, the low chromosphere is characterized by weak He I lines and strong metal lines (from Ti II, Fe II, Ba II etc.). Higher up, around 1500 km, the metal lines attain the same strength as the He I lines, and in the high chromosphere, i.e., in spicules, the intensity of He I lines exceeds the metal-line intensity.

TABLE II.2

Waldmeier's classification

Class	Criteria	Objects
I	$I(b_3) < I(b_4)$	⎫
II	$I(b_3) = I(b_4)$	⎬ Prominences
III	$I(b_4) < I(b_3) \leq I(b_2)$	⎭
IV	$I(b_2) < I(b_3) \leq I(b_1)$	⎫ Flares
V	$I(b_3) > I(b_1)$	⎭

The first modern spectral classification is due to Waldmeier (1949, 1951, 1961) who used the b lines of Mg I (b_1 at 5184 Å, b_2 at 5172 Å and b_4 at 5167 Å) and compared their intensities with that of b_3 of Fe II at 5168 Å (see Table II.2). Waldmeier found that while flares generally fall in classes IV and V (high excitation objects), prominences belong to class III, with a few in classes I and II. This classification distinguishes flares and prominences, but does not significantly subdivide prominences.

Another classification of solar atmospheric objects is due to Zirin and Tandberg-Hanssen (1960) (see Table II.3). This classification is based on the multi-component model, i.e., on the idea that flares and active prominences consist of different regions. Some regions are cool, less than 10^4 K, with strong H and metal-line emission, while the He emission is weak. Other regions are composed of a hotter plasma with temperature in excess of 10^4 K, giving rise to strong He and faint metal-line emission. A comparison between the classification lines gives a measure of which regions dominate a given object. As seen in Table II.3, such a comparison furnishes a convenient way of classifying different aspects of solar atmospheric activity and of com-

TABLE II.3

The Zirin–Tandberg-Hanssen classification

Class	Criteria	Objects
I	$I(\text{He\,\textsc{i}}, 4026) \ll I(\text{Sr\,\textsc{ii}}, 4078)$ $I(\text{He\,\textsc{i}}, 4713) \ll I(\text{Ti\,\textsc{ii}}, 4572)$ $I(\text{He\,\textsc{ii}}, 4686) \ll I(\text{He\,\textsc{i}}, 4713)$	Low Chromosphere
II	$I(\text{He\,\textsc{i}}, 4026) \approx I(\text{Sr\,\textsc{ii}}, 4078)$ $I(\text{He\,\textsc{i}}, 4713) \approx I(\text{Ti\,\textsc{ii}}, 4572)$ $I(\text{He\,\textsc{ii}}, 4686) \ll I(\text{He\,\textsc{i}}, 4713)$	Middle Chromosphere (around $h = 1500$ km) Quiescent Prominences
III	$I(\text{He\,\textsc{i}}, 4026) \gg I(\text{Sr\,\textsc{ii}}, 4078)$ $I(\text{He\,\textsc{i}}, 4713) \gg I(\text{Ti\,\textsc{ii}}, 4572)$ $I(\text{He\,\textsc{ii}}, 4686) \approx I(\text{He\,\textsc{i}}, 4713)$	High Chromosphere (spicules) Active Prominences Flares
IV	Presence of Fe x, 6374, Fe xiv, 5303, Ca xiii, 4086, or Ca xv, 5694	Coronal enhancements Coronal condensations

paring them with the chromosphere (the quiet Sun). Flares and active prominences fall in the same class III, characteristic of the high chromosphere, indicating that, spectroscopically, active prominences are more like flares than like quiescent prominences. The latter falls in class II and may be considered similar to the middle chromosphere in excitation. Extreme quiescent prominences may even border on class I.

The spectroscopic distinction between flares and active prominences, which is brought out in Waldmeier's classification, is not present in the original Zirin–Tandberg-Hanssen classification. To remedy this, the intensity of certain metal lines that behave differently in flares and in active prominences has been introduced as classification criteria (Tandberg-Hanssen, 1963). The distinction can be made by using lines of Fe II, which are strong in flares, and compare them with lines of Ba II and especially Ti II, which characterize spectra of prominences. Since the other classification lines in the easily observable blue-green part of the spectrum lie around 4600 Å, we choose the 4584 Å line of Fe II (multiplet 38) and the Ti II line at 4572 Å (multiplet 82) as our flare/active prominence discriminator. The ratio $M = I(\text{Fe\,\textsc{ii}}, 4584)/I(\text{Ti\,\textsc{ii}}, 4572)$, where $I(\text{Fe\,\textsc{ii}}, 4584)$ and $I(\text{Ti\,\textsc{ii}}, 4572)$ are the intensities of the two lines respectively, will then classify an active limb event as a prominence when $M < 1$ and as a flare when $M \geq 1$.

(b) *Continuum Emission*

Even though line emission dominates most prominence spectra, one often sees distinct continuous emission from loops and other active prominences, and in spectra of flares. Such emission can be due to free-free, or free-bound transitions, to H^- emission, as well as to synchrotron emission or scattering. In prominences, the dominant source is electron scattering of photospheric radiation. We return to the role of other continua in Section 2.1.5b.

The cross section per electron for electron (Thomson) scattering is $\sigma_e = 6 \cdot 6 \times 10^{-25}\,\text{cm}^2$ and the scattered intensity is

$$I_{sc} = \sigma_e W N_e I_{ph} \text{ erg s}^{-1} \Omega^{-1} \text{ cm}^{-1} , \qquad (\text{II}.19)$$

where I_{ph} is the intensity of the photospheric light per 1 cm of wavelength, and W is a geometrical dilution factor. Equation (II.19) can be used to find the electron density n_e in terms of the measured intensity I_{sc} and an estimated thickness X of the prominence with $N_e = n_e X$:

$$n_e = \frac{I_{sc}}{W \sigma_e X I_{ph}} . \qquad (\text{II}.20)$$

Ivanov-Kholodny (1959) deduced $n_e X = 10^{19}$ cm^{-2} from an observed value of $I_{sc}/I_{ph} = 10^{-6}$ and assuming $W = 0.2$. If the prominence is not homogeneous, but consists entirely of filamentary threads, the diameters of which have a characteristic dimension L, the local density will exceed the value given by Equation (II.20), which therefore gives a lower limit only. The question of the size of possible inhomogeneities, or of the value of the ratio

$$\alpha = \frac{L}{X} \qquad (\text{II}.21)$$

leads to one of the fundamental problems in contemporary prominence research.

2.1.5. HYDROGEN EMISSION

(a) Balmer Lines

From the observed halfwidth of Balmer lines we can assign upper limits to the kinetic temperature of the prominence plasma. In quiescent prominences, where the broadening velocity of H lines is about 10 km s^{-1}, the temperature must be less than, or at most equal to, 6000 K, and the uncertainty involved is quite small. Not so with density determinations. In principle, the electron density, n_e, can be determined from the observed Stark effect on the line profile. The wings of higher Balmer lines are widened in excess of a purely Doppler-broadened line, compare Equations (II.5) and (II.16). As we go to large values of the principal quantum number, n, the lines tend to coalesce and can no longer be resolved. The higher the electron density, the more pronounced is the effect, see Equation (II.15), and the fewer are the lines that can be distinguished in the spectra. This fact was used by Inglis and Teller (1939) who derived a relationship between the electron density and the quantum number, n_{max}, of the last resolvable Balmer line

$$\log n_e = 23.26 - 7.5 \log n_{max} .$$

Estimates based on this equation should be considered upper limits only, since the difficulties experienced in observing the higher Balmer lines generally prevent detection of lines corresponding to $n > 30$. Ivanov-Kholodny (1955, 1959) gives values of $n_e \approx 10^{12}$ cm^{-3}. He observed $n_{max} = 29$.

More recently, Jefferies and Orrall (1963) have used a refined theory of non-adiabatic effects of electron collisions (Griem, 1960, 1962) in studying Balmer lines up to $n = 36$, and conclude that $n_e = 5 \times 10^{10}$ to 10^{11} cm^{-3}.

Another way of using the Stark effect which is basically straightforward but which is difficult in praxis, is the actual observation of the deviation of the line profile in the wing from a Gaussian shape. Results based on this procedure should be considered uncertain, since the effect is discernible only in the far wing where the intensity of the line has fallen to less than 10% of the central intensity. Ivanov-Kholodny deduced from this method $n_e = 10^{12}$ cm^{-3}.

Hirayama (1963) re-evaluated the Stark effect in prominence Balmer lines, using Equations (II.3) and (II.12), i.e., he assumed a constant source function through the line-forming region and analyzed the observed profile in terms of the expression

$$\frac{I_\lambda}{I_0} = 1 - \exp\left[-\alpha_0 H(a, v)N\right]. \tag{II.22}$$

Under these assumptions Hirayama found n_e to range from 5×10^{10} to 5×10^{11} cm^{-3}, which agrees well with Jefferies and Orrall's results. Hence, we find for quiescent prominences electron densities of the order

$$n_e \approx 5 \times 10^{10} - 10^{11} \text{ cm}^{-3}. \tag{II.23}$$

If we combine the results of Equation (II.23) with the value for $N_e = n_e L = 10^{19}$ cm^{-2} given by Ivanov-Kholodny (1959), we find for the thickness of the fine-structure threads

$$L = 1000 - 2000 \text{ km} \tag{II.24}$$

or, by Equation (II.21)

$$\alpha \approx \tfrac{1}{6} - \tfrac{1}{3}. \tag{II.24'}$$

It should be noted that Ivanov-Kholodny used the value $n_e = 10^{12}$ cm^{-3} derived by him from his Stark effect analysis. He concluded that $L \approx 100$ km, $\alpha \approx 10^{-2}$, but since his value for the electron density is probably much too high, the resultant length-scale is too small. In any event, we find that, since the overall width of many prominences is of the order of 5000 to 10000 km, the prominences material is concentrated in fine threads that make up only a comparatively small fraction of the total volume generally called a prominence. We will return to this in Chapter III.

(b) *Continuum Emission*

Important information on both temperature and density may be obtained from a study of the H continua. The assumption is generally made (Zanstra, 1950; Redman and Zanstra, 1952; Orrall and Athay, 1957; Jefferies and Orrall, 1961a, 1961b) that the continuous emission observed in prominences is due entirely to H and to electron scattering of photospheric radiation (Section 2.1.4b). Under such conditions it is convenient to analyze the continuum emission in terms of the ratio R (Orrall and Athay, 1957; Jefferies and Orrall, 1961a, 1961b)

$$R = \frac{\text{Balmer continuum at 3646 Å}}{\text{All other continua at 3646 Å}}$$

or

$$R = \frac{\varepsilon_{\infty,2}}{\varepsilon_{ff} + \sum\limits_{n=3}^{\infty} \varepsilon_{\infty,n} + \varepsilon_{sc} + \varepsilon_{H^-}}, \tag{II.25}$$

where the free-free emission is given by

$$\varepsilon_{\text{ff}} = 1{\cdot}23 \times 10^{-19} n_{\text{p}} n_{\text{e}} T_{\text{e}}^{-1/2} \exp\left(-3{\cdot}94 \times 10^4 T_{\text{e}}\right) \tag{II.26}$$

and the free-bound emission by

$$\varepsilon_{\text{fb}} = 3{\cdot}89 \times 10^{-14} n_{\text{p}} n_{\text{e}} T_{\text{e}}^{-3/2} n^{-3} \exp\left[(15{\cdot}79 \times 10^4/n^2 T_{\text{e}}) - (3{\cdot}96 \times 10^4/T_{\text{e}})\right], \tag{II.27}$$

both in units of erg cm^{-3} s^{-1} Ω^{-1} $(\text{d}\lambda = 1)^{-1}$. In particular, we find for the Balmer continuum, $n = 2$,

$$\varepsilon_{\infty,2} = 4{\cdot}87 \times 10^{-15} n_{\text{p}} n_{\text{e}} T_{\text{e}}^{-3/2} . \tag{II.28}$$

The electron scattering term is

$$\varepsilon_{\text{sc}} = 2{\cdot}82 \times 10^{-11} n_{\text{e}} , \tag{II.29}$$

and the H^- emission

$$\varepsilon_{\text{H}^-} = B(T_{\text{e}}) \alpha_{\text{H}^-}(\lambda, T_{\text{e}}) k T_{\text{e}} n_{\text{e}} n_0 . \tag{II.30}$$

In the above formulae n_{p} is the proton number density, n_0 the number density of neutral hydrogen, $B(T_{\text{e}})$ the Planck function and α_{H^-} the absorption coefficient of the H^- ion, which is of importance only for temperatures $T_{\text{e}} < 10^4$ K, in which case (Chandrasekhar and Breen, 1946)

$$\alpha_{\text{H}^-} = 3{\cdot}80 \times 10^{-28} \exp\left(2{\cdot}58 \times 10^4/T_{\text{e}}\right) . \tag{II.31}$$

When we substitute Equations (II.26) through (II.31) in Equation (II.25), we find a quadratic expression for n_{e}. For T_{e} smaller than a certain T_{min}, no real value for n_{e} exists, and T_{min} depends on n_0. As n_{e} increases, R will approach a limit R_∞, thereby establishing an upper limit T_{max} on T_{e}.

To actually observe the ratio R, Jefferies and Orrall (1961b) showed that if one considers the emission at 3639 and at 3699 Å, R may be written in the form

$$R = \frac{\varepsilon(3639) - \varepsilon(3699)}{\varepsilon(3699)}, \tag{II.32}$$

since $\varepsilon(3639) = \varepsilon_{\text{ff}} + \sum_{m=3}^{\infty} \varepsilon_{\infty,n} + \varepsilon_{\text{sc}} + \varepsilon_{\text{H}^-}$, and $\varepsilon(3639) - \varepsilon(3699) = \varepsilon_{\infty,2}$.

The observed intensities at these wavelengths, i.e., $I(3639)$ and $I(3699)$, in general are complicated integrals over the emissions ε. However, if we assume that the emissions can be considered constant throughout the prominence, R is given simply by

$$R = \frac{I(3639) - I(3699)}{I(3699)} . \tag{II.33}$$

In quiescent prominences one finds $R \approx 5$. Jefferies and Orrall have studied the R ratio in different objects, and find solution sets $\{T_{\text{e}}, n_{\text{e}}\}$ for a quiescent prominence and for a flare-like loop of $\{12000$ K, 8×10^{10} $\text{cm}^{-3}\}$ and $\{15000$ K, 2×10^{11} $\text{cm}^{-3}\}$ respectively. While the density values seem reasonable, the temperature quoted for the quiescent prominence is uncomfortably high. On the other hand, Kawaguchi (1965) has shown that a value $R = 5$ in a quiescent prominence is consistent with the previously adopted temperature ≈ 6000 K, leading to a solution set $\{6000$ K, $5 \times 10^{10} \text{cm}^{-3}\}$.

2.2. Motions

We have referred to the motions of prominences in the Introduction, and the nature and degree of motions have been used in several of the classification schemes. We have seen that it is the speed of the prominence material that determines what we describe as active or quiescent objects. In this section we shall investigate more closely the nature of motions in different types of prominences.

2.2.1. QUIESCENT PROMINENCES

Under sufficiently good seeing conditions, mass motions can be observed in all quiescent prominences. This means that even though the overall shape of such prominences remains essentially unchanged over long periods of time, the material at any one point in the filament is in motion. We shall first discuss quiescent prominences well removed from any sunspot and where there is no interaction between the ends of the prominence and active regions. On high resolution Hα photographs one observes that the material is concentrated in thin ropes of diameter less than 300 km (Dunn, 1960), and from movies one gets the impression that the material is slowly streaming down these more or less vertical ropes (see also Dunn, 1965). Superimposed on this basically vertical gross-velocity, V, different knots of threads of the prominence plasma exhibit a random motion, v. According to Pettit (1932), Newton (1934), and Engvold (1971) for quiescent prominences, we have approximately:

$$v = 5-10 \text{ km s}^{-1} . \tag{II.34}$$

During activations of quiescent prominences (see Section 2.2.2) velocities $v=30$ to 50 km s^{-1} and even more can be observed (Larmore, 1953). However, the thermodynamic velocity parameter (the line-broadening velocity ξ_0) is always much smaller, again of the order of 5 to 10 km s^{-1}.

The gross velocity V is significantly smaller than the random velocity v. Nevertheless, the resulting mass loss from the prominence to the low chromosphere and photosphere is large enough to destroy a quiescent prominence in a time small compared to its lifetime. We are therefore forced to consider a dynamic model for even the most quiescent of prominences. Such models demand that material is continuously being fed into the prominence, albeit at a rate so slow that spectroscopic observations may hardly be affected. But physically, a quasi-dynamic model is implied. We return to the ramifications of this in Chapters III and IV.

Not all quiescent prominences are as quiescent as the ones described by Dunn's data. Initially, quiescent prominences form in pre-existing plage areas (d'Azambuja and d'Azambuja, 1948) that may contain sunspots. As a result, we often observe a quiescent prominence one end of which interacts with a sunspot in such a way that material is streaming out of the prominence and down into the spot. In this case the gross velocity, V, is much higher than in the case of the slowly downward streaming motions treated above. Due to the mass losses involved, the quiescent prominences that interact with active regions strongly call for a dynamic model.

In Section 2.2.3 we shall treat active region filaments that show streaming motions along most of its length, not only out of the one end. They are not to be considered quiescent prominences, but there may well be a generic relationship between the two types, a question that will be discussed in Chapter IV.

2.2.2. ACTIVATED QUIESCENT PROMINENCES

At times quiescent prominences are subjected to external disturbances that result in motions of the prominence plasma. The severity of the perturbations may range from a slight temporary 'activation', manifested by increased internal motions, to disturbances of such magnitude that the prominence is completely destroyed. Thus, we may distinguish between several degrees of violence in the observed motions, and we shall now discuss several cases. The disturbing agents generally emanate from flares, in some instances from growing sunspots; their nature will be discussed in Chapter VI.

(a) *Increased Internal Motions*

This type of activation – characterized by chaotic motions of knots and other fine-structure details of the different parts of the prominence plasma – leads to velocities $v = 30$ to 50 km s^{-1} and is often accompanied by a slow rising motion of the whole prominence (Bruzek, 1969b). After the activation the prominence may return to its original quiescent state. However, in other cases the increased internal motion is an indication that a major disturbance is in progress. In other extreme cases the activation may impart very strong internal motions to the plasma, up to 300 km s^{-1}, and then die down again (Valníček, 1968). Such violent activities do not destroy the prominence, but significant changes occur. Of particular interest is the observation of a de-twisting motion of the interwoven helices that make up the prominence. We shall return to the importance of helical structure of prominences later.

(b) *Winking Filaments*

The motion to be described here is due to the fact that the filament as a whole is subjected to more or less vertical oscillations. This leads to a prominence spectrum whose lines are Doppler shifted alternately to the red and violet as the filament is pushed up or down in the corona. As a consequence the image of the prominence is shifted alternately out of and into the passband of narrow-band Hα filters, and gives the 'winking' impression. A winking filament generally shows an initial downward motion followed by from one to four damped oscillations. The phenomenon has been known since the spectrohelioscopic observations by Greaves, Newton and Jackson (Dyson, 1930) and Newton (1935). Ramsey and Smith (1966) observed a filament to wink four times in three days, and each time the oscillation frequency, v_{osc}, was essentially the same, viz:

$$v_{osc} \approx 10^{-3} \text{ s}^{-1} . \tag{II.35}$$

As we shall see later the winking is caused by an explosive perturbation emanating from a flare, but in these cases the activation is a mild one, never causing serious disruption of the prominence.

(c) *Disparitions Brusques*

A gross velocity of particular interest results when a quiescent prominence undergoes a sudden disappearance due to an ascending motion. To distinguish it from other types of disappearances, we employ the often-used French term, disparition brusque. Also in these cases the cause generally is a flare-induced activation, and here the perturbation has a profound influence on the stability of the filament. The correlation with flares has been studied by Bruzek (1951), Øhman and Øhman (1953), Øhman et al. (1962), and Westin and Liszka (1970). Prior to a disparition brusque the prominence material exhibits increased random velocities, $v \approx 30$ to 50 km s^{-1} (see Section (a)), and then the whole prominence, or most of it, starts to ascend with increasing velocity V. This gross velocity attains values of several hundred km s^{-1}. The well publicized observation of the nearly classical event of June 4, 1946 is still unsurpassed in beauty, and a sequence of photographs in Hα of this prominence may still offer the best illustration of the disparition brusque phenomenon (see Figure II.2). As the disparition brusque phase commences the quiescent prominence is ejected into the corona with steadily increasing velocity, resulting in a velocity curve (Valniček, 1964) as exemplified by curve II in Figure II.3. The prominence material continues to be acted upon by a force, and may reach velocities in excess of the escape velocity, V_{esc}, (Menzel et al., 1956; Bjerke, 1961). The value of V_{esc} falls from 618 km s^{-1} at the photosphere to about 400 km s^{-1} at a height of $h = 100000$ km in the corona, according to the expression

$$V_{esc} = \sqrt{\frac{2GM_\odot}{h}}, \tag{II.36}$$

which is deduced from the vis viva (conservation of energy) equation, and where $M_\odot$ is the mass of the Sun and G the gravitational constant. Valniček designates as group II those prominences that show velocity curves like curve II in Figure II.3, i.e. prominences that move out with increasing gross velocity. Quiescent prominences during their disparition brusque phase furnish the most important examples of this group.

In addition to the outward velocity of the erupting material, the disparition brusque phase is characterized also by a most interesting spiralling motion of the prominence material (Severny and Khoklova, 1953; Zirin, 1968; Malville, 1969; Slonim, 1969; Anzer and Tandberg-Hanssen, 1970). As we shall see in Chapter VI the eruption seems to be due to the action of the prominence-supporting magnetic field. The spiralling, therefore, reveals some of the inherent nature of these prominences, being due to the interaction of the rising magnetic field with the prominence plasma.

According to d'Azambuja and d'Azambuja (1948) nearly one-half of all low-latitude filaments are seen to suffer a disparition brusque and disappear temporarily at least once. Consequently, the disparition brusque phase must be considered a 'normal' experience for quiescent prominences, probably all go through it. The disappearance lasts from a day up to a few weeks, after which the prominence reforms in two-thirds of all cases. It is important to notice that when the filament reforms, it appears in very nearly the same shape as before.

(d) Sinking and Shrinking Filaments

The last types of activations that we shall discuss here lead, like the disparition brusque, to the disappearance of the whole prominence. But instead of ascending and disappearing, and then generally reforming, the sinking and shrinking filaments are subjected to an activation that may look less spectacular, but that may have a more drastic outcome: these filaments disappear, generally never to reform. In these cases

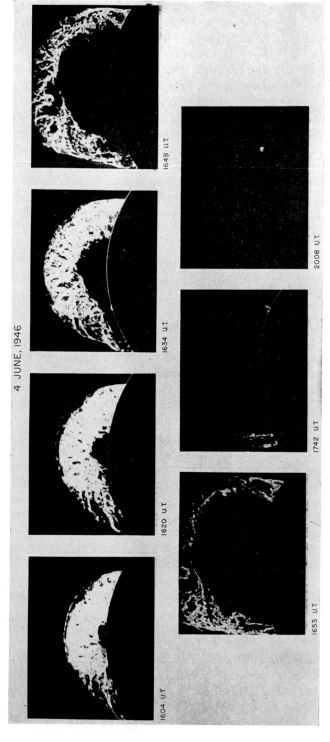

Fig. II.2. The disparition brusque phase of the prominence of June 4, 1946.

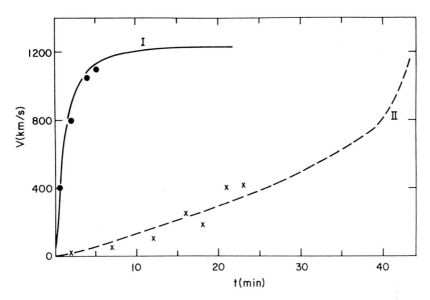

Fig. II.3. Velocity curves (after Valniček, 1964) for ejections, curve I, and activated quiescent prominences, disparition brusque phase, curve II. Dots refer to the prototype spray of Febr. 10, 1956 (see
for instance, E. v. P. Smith, 1968). Crosses represent data for small eruptive prominence (Rosseland
and Tandberg-Hanssen, 1957).

we witness the whole prominence disintegrate by gradual dissolution, by falling into
the Sun, or both. Again, the gross velocity, V, is considerable, easily attaining 100
km s^{-1} (see Kiepenheuer, 1953a).

2.2.3. ACTIVE REGION FILAMENTS

The active region filaments (the sunspot filaments in Zirin's classification) are characterized by motion along the filament. In high resolution movies one can see a continuous
flow of material along the axis of most such prominences (S. F. Smith, 1968). This
flow is one of the strongest descriptive differences between active and quiescent
filaments. In the most active of quiescent prominences there is hardly any mass motion
observed along the long axis of the filament. As we go toward more active conditions
this situation changes. For example, if one end of the quiescent filament is near an
active region like a sunspot, we may observe significant motion along part of the
filament and out of one end into the sunspot. Finally, in active filaments the whole
prominence plasma is continuously flowing along the prominence axis, a situation
that demands a completely dynamic model. Also, this discussion points to the possibility that quiescent and active filaments somehow may be related. We return to this
question in Chapter VI. Consult also Figure II.11.

2.2.4. ARCH FILAMENT SYSTEMS

Analysis of the filtergrams and Hα spectra show that matter streams down both legs
of arch filaments (Bruzek, 1969a). The gross velocity is $V \approx 50$ km s^{-1}. In addition,
the tops of the arches are observed to ascend at velocities up to 10 km s^{-1}. We shall

see in Chapter III that these arches may be considered magnetic flux tubes that rise in the atmosphere.

2.2.5. EJECTIONS

Some types of prominences owe their existence to the fact that at times material is ejected from certain active regions. When this material is seen projected on the plane of the sky above the limb, we observe an ejected prominence. The active region involved is nearly always flare-producing, and with increasing importance of the triggering explosions we denote the ejections as 'surges', 'sprays' and 'fast ejections' (Bruzek, 1969b). When the ejecta are observed against the disk of the Sun, we may see them either in absorption or in emission, giving rise to dark or bright surges and sprays.

(a) Surges

In using the above nomenclature we have followed Giovanelli and McCabe (1958) and Bruzek (1969b) who describe surges as straight or slightly curved spikes which grow out rapidly of a small, round, luminous mound, associated with, or being, a flare (Gopasyuk and Ogir, 1963). Surges are shot out at velocities of 100 to 200 km s^{-1}, typically last 10 to 20 min, and reach heights in the corona of 100000 to 200000 km. As the material is ejected, it often performs a spiralling motion on its way out. After the maximum height is reached, the material falls back into the chromosphere along the trajectory of its ascent. This falling back often seems to trigger a new surge (Zirin and Werner, 1967). Surges show a strong recurrence tendency (McMath and Pettit, 1938; Lodén, 1958).

(b) Sprays

These prominences are invariably ejected from flares during their expanding phase (Warwick, 1957; Švestka, 1962; E. v. P. Smith, 1968; Bruzek, 1969b; McCabe, 1971). One sees on the solar limb an exploding flare mound that ejects material at velocities, V, generally exceeding the velocity of escape, $V > V_{esc}$, but ranging from maybe 200 to 2000 km s^{-1}. Characteristically, the explosion is so violent that the material is not contained, as in the case of surges, but flies out, fragmented into 'bits and pieces'.

Some authors prefer to reserve the name spray for an ejection whose velocity exceeds V_{esc} (Smith and Smith, 1963). This is an important characteristic from the point of view of the influence of the ejection on interplanetary space. However, from the point of view of the physics of the prominence it is probably not crucial whether or not the escape velocity is reached. We consider as physically important the fact that the explosion is so violent that the ejected material cannot be contained (by the magnetic fields) and we thereby distinguish sprays from surges.

The reality of the observed clumpiness or fragmentation of the ejected matter is not obvious. To a large extent it may be due to the narrowness of the instrumental passband used. With broadband observations at times one can see massive prominences where the spray appears fragmented as observed with narrow passband filters (Øhman et al., 1967). Again, we take the fragmentation as an indication of the violence of the event, observable under certain types of observations.

(c) *Fast Ejections*

Bruzek (1969b) has called attention to some fast ejections which seem to form a class of ejecta distinguishable from sprays. These very fast ejections consist of a compact portion of a flare which is ejected as a whole, and is not fragmented. The velocities involved lie in the upper range of the spray velocities, and these ejections will therefore influence conditions in interplanetary space.

The velocity curve (Valniček, 1964) for sprays and fast ejections is shown schematically as curve I in Figure II.3, and is seen to be quite different from that of eruptions. The ejections reach velocities of 500 to 2000 km s^{-1} in a few minutes, and are thence not sustained by a continuing acceleration. We are witnessing a physical process, some kind of explosion, profoundly different in nature from the activation of a quiescent prominence which leads to a disparition brusque.

2.2.6. LOOPS AND CORONAL RAIN

Some of the most beautiful examples of gracious motion of material in the corona are provided by loop prominences and coronal rain, types A in the Menzel–Evans classification. In these objects matter is observed to flow down along curved trajectories into active regions, seemingly after condensing out of the corona, often at great heights.

(a) *Loops*

Loop prominence systems develop out of a bright flare mound which in the beginning grows at a rate of up to 20 km s^{-1} (Dodson-Prince, 1961; Bruzek, 1964a, 1969b). The mound divides into a number of loops in which the material is streaming down both legs. The streaming motion follows a single arc, one does not observe the spiralling motion often seen in surges and in erupting quiescent prominences. During the next several hours the loop system expands and reaches, typically, a height of 50 000 km. It is important to note that the individual loops do not seem to grow or expand much, rather the system expands by generating higher and higher loops, while the lower ones fade away. The resulting apparent velocity of expansion is quite small, ≈ 5 km s^{-1}. The real, downward flow motion in the two legs of the loop is considerably faster; at impact with the chromosphere or photosphere it is of the order of the velocity of free fall, i.e.,

$$V \approx V_g = a_\odot t \approx 130\text{–}160 \text{ km s}^{-1}, \qquad (\text{II}.37)$$

for 8 to 10 min. fall.

(b) *Coronal Rain*

Material is often seen to rain down into active regions from coronal clouds suspended in the corona. The trajectories followed by the streaming material generally are strongly curved, apparently following the lines of force of coronal magnetic fields (see Figure II.4). The velocities along these paths are comparable to the flow velocities observed in loop prominences, and it is at times difficult to distinguish between the two types.

AUGUST 12, 1951

Fig. II.4. Coronal cloud with material streaming down in curved trajectories toward active center.

2.3. Magnetic Fields

2.3.1. Historical Remarks

The single, physically most important, parameter to study in prominences may be the magnetic field. Shapes, motions, and in fact the very existence of prominences depend on the nature of the magnetic field threading the prominence plasma. It is therefore natural that attempts have been made to infer the configuration and/or strength of the fields involved from studies of the shapes and motions of active prominences (Correll *et al.*, 1956; Idlis *et al.*, 1956; Warwick, 1957; Correll and Roberts, 1958; Bumba and Kleczek, 1961; Kleczek 1962, 1968; Billings, 1966; Hyder, 1966). Babcock and Babcock (1955) pointed out that quiescent prominences seen on the disk lie along the neutral line between the two opposite polarities in bipolar magnetic regions (see also Stepanov (1958) and Howard (1959)). Hyder (1964a, b) used the theory of resonance polarization and impact polarization (Warwick and Hyder, 1964) to estimate the longitudinal magnetic field in prominences from studies of the observed linear polarization of prominence emission lines.

When solar magnetographs were adapted to observe the Zeeman effect in spectral lines from prominences, actual measurements of the longitudinal magnetic field became possible (Zirin, 1961; Zirin and Severny, 1961; Ioschpa, 1962, 1963; Rust 1966, 1967; Harvey and Tandberg-Hanssen, 1968; Malville, 1968; Harvey, 1969; Smolkov, 1970; Tandberg-Hanssen, 1970). The measurements rely on observations of the circularly polarized σ-components of Zeeman-affected spectral lines. In this way one gets information on the longitudinal component only of the magnetic field. Observations of linear polarization in Zeeman-affected lines leading to deduced transverse magnetic fields have not been accomplished for prominences, even though several attempts are presently being made.

2.3.2. THE ZEEMAN EFFECT

In the simplest case a single spectral line with wavelength λ_0 will be split into three components, a π-component which is not shifted in wavelength relative to the un-affected line and which is linearly polarized, and the two σ-components. The σ-components are shifted in wavelength a distance $\Delta\lambda_B$ from λ_0, one to the red and the other to the blue of the π-component. The shift is given by

$$|\Delta\lambda_B| = \frac{c}{4\pi m_e c} \lambda_0^2 gB = 4\cdot67 \times 10^{-13}\lambda_0^2 gB \tag{II.38}$$

where λ is in Ångstrom and the strength of the magnetic field B is in Gauss. The Landé g-factor determines the splitting of the line through the magnetic moment, μ_B, of the atom in the direction of the field B

$$\mu_B = \frac{eh}{4\pi m_e c} Mg \ .$$

M, the magnetic quantum number, follows the rule $\Delta M = \pm 1$. When $\Delta M = 0$ we observe the unshifted π-component.

When the magnetic field is strong enough, the two σ-components are separated and one can measure the displacement $\Delta\lambda_B$ on a photographic plate of the spectrum. However, this requires fields of several thousand Gauss, and the method therefore can be used only in sunspots. In prominences, one makes use of the other property of the Zeeman effect on the lines, namely the different sense of polarization.

For observations outside the limb of the Sun, i.e., in prominences and in the corona, we see emission lines, and the presence of magnetic fields leads to the Zeeman effect just described. When we observe the disk of the Sun, i.e., plages, sunspots, etc., we see the spectral lines in absorption (Fraunhofer lines), and the presence of a magnetic field gives rise to the so-called inverse Zeeman effect. One must then consider the more complicated case of a gas which both emits and absorbs radiation. The absorption properties may be described by introducing three selective absorption coefficients, k_π for plane polarized light in the direction of the magnetic field, $k_{\sigma,1}$ for left-circularly polarized light, and $k_{\sigma,r}$ for right-circulary polarized light (Unno, 1956). Similarly, selective emission coefficients, $j_\pi, j_{\sigma,1}$ and $j_{\sigma,r}$ may be used to account for the emission properties. We shall, in the following, mainly treat the emission Zeeman effect.

There are several ways by which one may describe the state of polarization of a light beam (see for instance, Shurcliff, 1962; Collett, 1968; Harvey, 1969). We may represent the beam by a pair of orthogonal plane waves

$$\begin{aligned} E_x(t) &= E_x^0(t) \cos\left[\omega t + \delta_x(t)\right], \\ E_y(t) &= E_y^0(t) \sin\left[\omega t + \delta_y(t)\right], \end{aligned} \tag{II.39}$$

where $E_{x,y}^0(t)$ are the instantaneous amplitudes of the electric field, ω is the frequency and $\delta_{x,y}$ are the phase factors. From Equation (II.39) we can find the expression for the polarization ellipse, which is a familiar way of describing a beam of elliptically polarized light. Instead, we may describe it by the completely equivalent 4 Stokes parameters (Chandrasekhar, 1960)

$$
\begin{aligned}
I &= E_x^{02} + E_y^{02}, \\
Q &= E_x^{02} - E_y^{02}, \\
U &= 2E_x^0 E_y^0 \cos \delta, \\
V &= 2E_x^0 E_y^0 \sin \delta,
\end{aligned}
\tag{II.40}
$$

where $\delta = \delta_y - \delta_x$.

The parameter I is the total intensity of the radiation and Q, U and V describe the state of polarization. Furthermore, we see that $I^2 = Q^2 + U^2 + V^2$. In other words, the Stokes parameters are the observables of the polarization ellipse, or of the light beam. If the beam is incompletely polarized, then the intensity referred to above represents the polarized part, I_p, only, so that I_p/I gives the degree of polarization,

$$
P \equiv \frac{I_p}{I} = \frac{\sqrt{Q^2 + U^2 + V^2}}{I}.
$$

It is possible to write the Stokes parameters as four elements of a single column matrix, $\{I, Q, U, V\}$, and the intensities of the Zeeman components are then

$$
\begin{aligned}
I_\pi &= j_\pi \{\sin^2 \gamma, \sin^2 \gamma, 0, 0\}, \\
I_{\sigma, r} &= j_{\sigma, r} \{\tfrac{1}{2}(1 + \cos^2 \gamma), -\tfrac{1}{2} \sin^2 \gamma, 0, \cos \gamma\}, \\
I_{\sigma, 1} &= j_{\sigma, 1} \{\tfrac{1}{2}(1 + \cos^2 \gamma), -\tfrac{1}{2} \sin^2 \gamma, 0, -\cos \gamma\},
\end{aligned}
\tag{II.41}
$$

where γ is the angle between the line of sight and the direction of the magnetic field, and the emission coefficients have the form

$$
j_\pi = j(v), \text{ (i.e. identical to the coefficient in the absence of a field)},
$$

$$
j_{\sigma, r} = j(v + v_B) = j(v) + v_B \frac{dj}{dv} + \cdots \quad \text{(Taylor series for small B)}
$$

$$
j_{\sigma, 1} = j(v - v_B) = j(v) - v_B \frac{dj}{dv} + \cdots.
\tag{II.42}
$$

In the case of absorption lines (the inverse Zeeman effect) completely analogous expressions are used for the absorption coefficients k_π, $k_{\sigma, r}$ and $k_{\sigma, 1}$.

In the case of the simple Zeeman triplet we find the following relationship between the direction of the magnetic field and the observed polarization of the line components (Equation (II.41)). When we observe parallel to the field, the so-called longitudinal case, $\gamma = 0$, we see only the two σ-components, circularly polarized in opposite directions, σ_R and σ_V, see Figure II.5. If the line is in emission, and the field is pointing toward the observer, σ_R will be right circularly and σ_V left circularly polarized. For an absorption line, and with the same orientation of the field, the σ_R component will be left circularly and σ_V right circularly polarized. In the other limiting case, when the

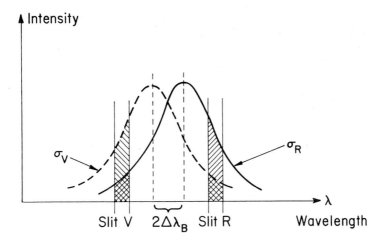

Fig. II.5. Zeeman components of a simple Zeeman triplet.

line-of-sight is perpendicular to the magnetic field, $\gamma = 90°$, the transverse case, the π-component is twice as strong as a σ-component, and is linearly polarized, parallel to the magnetic field in emission Zeeman effect, and perpendicular to the field in the inverse Zeeman effect. In this case, also the σ-components are plane polarized. For all intermediate cases, i.e., in all practical application nearly, we see elliptically polarized σ-components, while the π-component remains linearly polarized.

It is difficult to find suitable solar spectral lines that show the simple Zeeman splitting. Most lines actually used in magnetograph work show the so-called anomalous Zeeman effect, i.e., they split up into an often large number of σ- and π-components. The overlapping of these differently polarized components causes serious difficulties in interpreting the observed polarization in terms of a magnetic field. One should therefore divide the problem of measuring solar magnetic fields into its two parts, viz. (1) Determination of the state of polarization in certain selected spectral lines, and (2) Interpretation of the observed polarization in terms of a magnetic field. While the first part of the problem is mainly a question of techniques and is well understood (albeit being difficult), the second part relies on our understanding of the formation of spectral lines in the presence of a magnetic field, and a general theory is not available. The problem has been solved for special cases (Unno, 1956; Stepanov, 1958, 1960; Michard, 1961; Rachovsky, 1961, 1963, 1967; Obridko, 1965, 1968).

2.3.3. MAGNETOGRAPH OBSERVATIONS OF PROMINENCES

The basic principle of the longitudinal polarimeter (magnetograph) consists in alternately blocking out one of the two σ-components with appropriate polarizing optics, and measuring the difference in signal from the two components by using two phototubes, one fed by light from an entrance slit in the violet wing of the σ_V-component, slit V, the other by light transmitted through slit R in the red wing of the σ_R-component (see Figure II.5). For details on how magnetographs work the reader is referred to Beckers (1968b) and Harvey (1969), whose derivations we will follow below.

To find how the signal from the phototubes is related to the magnetic field one needs a theory for the formation of the line. In the simple case of a uniform magnetic field and an optically thin line, it can be shown that the resulting intensities of the Zeeman components (Equations (II.41)) lead to the following Stokes parameters

$$I = j_\pi \sin^2 \gamma + \tfrac{1}{2}(j_{\sigma,1} + j_{\sigma,r})(1 + \cos^2 \gamma) \,,$$

$$Q = j_\pi \sin^2 \gamma - \tfrac{1}{2}(j_{\sigma,1} + j_{\sigma,r}) \sin^2 \gamma \,,$$

$$U = 0 \,,$$

$$V = (j_{\sigma,r} - j_{\sigma,1}) \cos \gamma \quad \left(\approx 2B \frac{dj_\nu}{d\nu} \cos \gamma \right) . \qquad (\text{II.43})$$

In the case of the longitudinal polarimeter this light is transmitted through a circular polarizer, i.e., through a $\lambda/4$ plate followed by a linear polarizer oriented with its transmission axis $45°$ with respect to the fast axis of the $\lambda/4$ plate. The transmitted intensity is then

$$I_t(\|) = \tfrac{1}{2}[j_\pi \sin^2 \gamma + \tfrac{1}{2}(j_{\sigma,1} + j_{\sigma,r})(1 + \cos^2 \gamma) \pm (j_{\sigma,1} - j_{\sigma,r}) \cos \gamma] \,.$$

Using the expression (II.42) for the emission coefficients j and retaining only the first term of the Taylor expansion, we find

$$I_t(\|) \approx j_\nu \pm \nu_B \cos \gamma \frac{dj_\nu}{d\nu} = j_\nu \pm \text{const. } B \cos \gamma \frac{dj_\nu}{d\nu} \,.$$

With the help of the two slits of the polarimeter the phototubes measure the intensity at a point in the line profile with alternately the σ_R and the σ_v component blocked out. By taking the difference one obtains a signal $s(\|) = I_{t,R}(\|) - I_{t,v}(\|)$, or

$$s(\|) = \text{const. } B \cos \gamma \frac{dj_\nu}{d\nu} \,, \qquad (\text{II.44})$$

i.e., the resultant phototube signal varies as $B \cos \gamma$. We see that for small magnetic fields the $s(\|)$ signal is just the Stokes parameter V of Equation (II.43). To derive $B \cos \gamma$ from V, one must determine $dj_\nu/d\nu$. This is done with the help of the Stokes parameter I, measuring the slope of the line profile as well as its intensity.

The problems involved in measuring polarization are vastly increased when we want to know the complete state of polarization, which is needed to find the vector magnetic field. The transverse component of the field is associated with the Stokes parameter giving linear polarization, i.e., with the π-component of the Zeeman-affected spectral line. In this transverse case the light beam of Equation (II.43) is transmitted through a linear polarizer and the transmission axis is switched between a direction parallel to and a direction perpendicular to the magnetic field. One can then show that the transmitted intensity is given by

$$I_t(\perp) = \tfrac{1}{2}[j_\pi(\sin^2 \gamma \pm \sin^2 \gamma) + \tfrac{1}{2}(j_{\sigma,1} + j_{\sigma,r})(1 + \cos^2 \gamma) \pm (j_{\sigma,1} + j_{\sigma,r}) \sin^2 \gamma] \,.$$

Again, using Equations (II.42) with two terms of the Taylor expansion retained, we derive the following expression for the signal due to the intensity difference $I_{t,R}(\perp) - I_{t,v}(\perp)$ after transmission through the alternating linear polarizer

$$s(\perp) = \text{const. } B^2 \sin^2 \gamma \frac{d^2 j_\nu}{d\nu^2} \,. \qquad (\text{II.45})$$

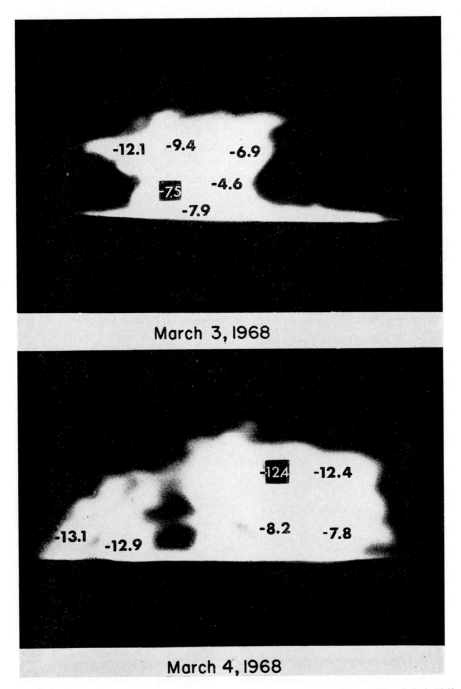

Fig. II.6. The longitudinal magnetic field observed in a quiescent prominence, March 3–4, 1968.

We see that measuring the linear polarization is associated with measuring the Stokes parameter Q, which varies as the square of the transverse field. Comparing Equations (II.44) and (II.45), we notice that below a certain field strength it is inherently more difficult to measure the transverse field than the longitudinal field. An additional difficulty is encountered in measuring the transverse field in that the direction of polarization is not known *a priori*. There is an ambiguity of 180° in the direction of the transverse component, and it is therefore necessary to measure both the Q and the U Stokes parameters to define the transverse field.

If in addition to measuring the Stokes parameters Q and U, also the parameter V is included, we have a complete Stokes parameter polarimeter, giving complete information as to the state of polarization of a spectral line. Provided we can interpret the observed polarization in terms of a magnetic field, we have a full-fledged vector magnetograph.

2.3.4. QUIESCENT PROMINENCES

(a) *Limb observations*

Comprehensive studies of magnetic fields in quiescent prominences have been published by Rust (1966), Smolkov (1966, 1970), Harvey (1969), and Tandberg-Hanssen (1970). When observed above the limb the longitudinal field, $B_{\parallel}$, of the prominence is measured through an aperture of the order of 10 by 10^{11}. Averaging over such an area

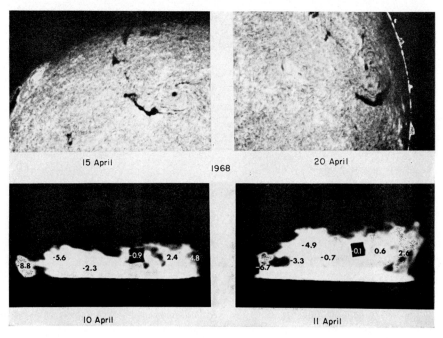

Fig. II.7. Quiescent prominence showing longitudinal magnetic field of both polarities, probably due to geometrical effects.

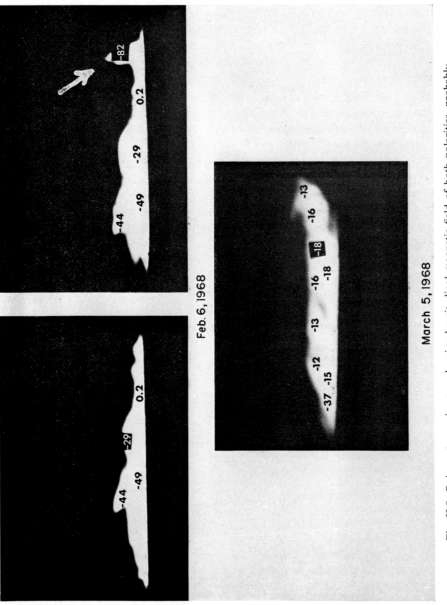

Fig. II.8. Quiescent prominences showing longitudinal magnetic field of both polarities, probably due to surge activity at one extremity.

we normally find values for $B_{\parallel}$ ranging from no observable field (i.e. $B_{\parallel} <$ about $0\cdot5$ G) to 30 or 40 G.

(α) *Hα-observations.* In quiescent prominences a series of measurements of the field from different parts of the object is often possible (see Figure II.6). One finds then as a general rule that even though the measured field strength may change from place to place, the polarity $(-B_{\parallel}$ or $+B_{\parallel})$ does not change. Those relatively rare instances where both polarities are observed may be explained in terms of one of the two following situations.

(1) The prominence sheet does not form a plane, but is bent so that both sides of the prominence are visible simultaneously (see Figure II.7).
(2) Some part of the prominence is not quiescent, but shows for instance excessive streaming or surge activity (see Figure II.8).

Apart from these situations, we observe the magnetic field either as it enters one side of the prominence ($B_{\parallel}$ negative), or as it emerges ($B_{\parallel}$ positive), Figure II.8. Figure II.9 shows a histogram of the distribution of the mean longitudinal field, $\bar{B}_{\parallel}$, as observed with the Hα line, for 135 quiescent prominences recorded at Climax in 1968–69. The overall mean value for these prominences is $\langle\bar{B}_{\parallel}\rangle = 7\cdot3$ G (median value $6\cdot9$ G), and slightly more than 50% of the prominences have mean values satisfying the inequalities

$$3\,\text{G} \le \bar{B}_{\parallel} \le 8\,\text{G}. \tag{II.46}$$

Rust (1966) found $\langle\bar{B}_{\parallel}\rangle \approx 5$ G for data from 1965, and Harvey (1969) found $\langle\bar{B}_{\parallel}\rangle = 6\cdot6$ G. His data were obtained mainly in 1967. The differences between the values

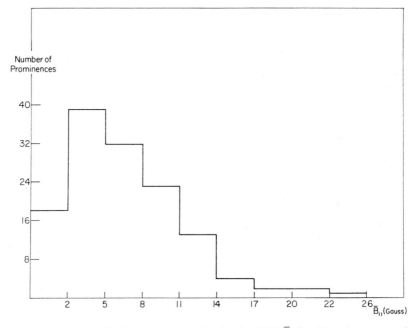

Fig. II.9. Histogram giving distribution of mean longitudinal field $\bar{B}_{\parallel}$ for 135 quiescent prominences.

Fig. II.10. Quiescent prominences showing longitudinal magnetic field increasing with height.

quoted may not be significant, merely due to selection effects. On the other hand, Harvey pointed to the possibility that the general level of prominence-supporting fields could have been greater in 1967 than closer to sunspot minimum in 1965, thereby explaining the difference between his and Rust's results. The 1968–69 Climax data support this point of view.

There is a general tendency for the observed field $B_\parallel$ to increase with height in the prominence, even though there are many prominences where this increase is masked by internal noise in the data (see Figure II.10).

(β) *Observations using several different spectral lines.* If we want information about the magnetic field using several different spectral lines, often, there is time only to make the measurements on a few selected spots in the prominence. Table II.4 summarizes pertinent data for a number of quiescent prominences whose magnetic field was determined by observing two or more of the H, He, or metal lines listed in Table II.5.

TABLE II.4

Observed longitudinal component of magnetic field, $B_{\parallel}$, in quiescent prominences

Date	Position angle	λ	$\dfrac{B_{\parallel}}{G}$	G	Object
June 26, 1968	250	D_3	12·6	±2	Small quiescent
		D_1	10	±4	
		Hα	9·0	±0·7	Same prominence June 25
Sept. 8, 1968	325	Hα	−0·1	±0·7	Small quiescent
		D_3	+0·3	±0·7	
Sept. 24, 1968	315	Hα	6	±0·5	Big quiescent
		D_3	5·5	±0·5	
		4471	7	±3	
Nov. 27, 1968	90	D_3	21	±0·5	Semi-quiescent
		4471	21	±4	
		D_1	18	±2	
		D_2	23	±2	
		b_1	15	±4	
Jan. 16, 1969	59	Hα	−16·6	±1	
		D_3	−18	±3	
		D_3	−15	±1	
		4471	−21	±3	
		D_1	−20	±2	
		D_2	−27	±2	
Jan. 21, 1969	86	Hα	−15·7	±0·5	Part of big quiescent
		4471	−12	±6	
		D_1	−15	±2	
		D_2	−14	±3	
April 5, 1969	310	Hα	−11·5 to −14·4	±0·5	Small quiescent
		D_3	−15·4	±1	
April 24, 1969	297	Hα	−29·4	±4	Semi-quiescent
		4471	−17	±5	
		D_3	−24	±1	
		D_1	−26	±2	
		D_2	−31	±2	
Aug. 16, 1969	130	Hα	5 to 8	±1	Small quiescent
		D_3	8 to 10	±1	
Jan. 4, 1970	75	D_3	13	±1	Big quiescent
		D_3	18	±2	
		D_3	14	±1	
		D_1	20	±4	
		D_1	15	±4	
		D_2	10	±8	
		D_2	13	±6	
		b_1	12	±2	
Jan. 6, 1970	100	Hα	−29	±0·5	
		D_3	−28	±1	

TABLE II.5

Parameters for spectral lines used to measure
longitudinal magnetic fields in prominences

Element	$\lambda(\overset{\circ}{A})$	Line Designation	$\bar{g}$
H I	6563	Hα	1·045
He I	5876	D$_3$	1·11
He I	4471	—	1·04
Na I	5889	D$_1$	1·33
Na I	5896	D$_2$	1·17
Mg I	5184	b$_1$	1·75

The most striking result of a perusal of Table II.4 is the impression that the magnetic field is the same whether observed with H, He, or metal lines, i.e.,

$$B_\parallel(H) \approx B_\parallel(D_3) \approx B_\parallel(D_1) \,. \tag{II.47}$$

(b) *Disk Observations*

When observed on the disk, quiescent filaments are found in extensive bipolar regions (remnants of old plages), and are seen to run along the neutral line of the longitudinal magnetic field of the region in question, see the schematic illustrations in Figure II.11. Notice that what we call active filaments are found in and near younger plages, and even though also they mostly run along the neutral line (Schoolman, 1969), these objects are not quiescent prominences. The nature of their magnetic field will be discussed in the next section.

2.3.5. ACTIVE PROMINENCES

(a) *Limb Observations*

The first measurements made of magnetic fields in prominences (Zirin, 1961; Zirin and Severny, 1961) pertained to active objects. Since then Ioshpa (1968) and Harvey (1969) have studied the fields in different types of active prominences. The observed values depend on the orientation of the prominence relative to the line of sight. This is especially true for surges, but any set of data will be strongly influenced by selection effects. For example, Harvey found a median value of 26 G for the active prominences he observed (about 200 single observations), while the median value for the 275 active prominences observed at Climax in 1968–69 (about 1000 single observations) is between 4 and 5 G. For this part of data $\langle \bar{B}_\parallel \rangle = 6\cdot8$ G, and only about 30% of the prominences have field satisfying the inequality

$$3\,\text{G} \le \bar{B}_\parallel \le 8\,\text{G} \,. \tag{II.48}$$

Comparing these results with the corresponding data for quiescent prominences (see discussion leading to expression (II.46)), we find that while the overall mean field for active prominences is not very different from that of quiescent objects, the spread is much greater. There are many active prominences whose longitudinal magnetic field is 2 G or less; in the Climax sample from 1968–69, 18%, versus 13% for quiescent

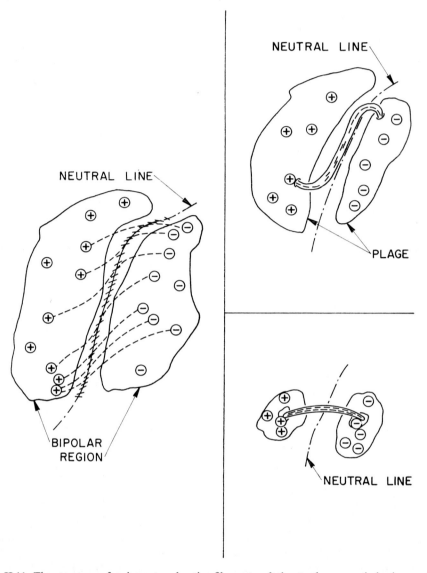

Fig. II.11. The structure of quiescent and active filaments relative to the magnetic background in which they are found (schematic). Left: quiescent prominence; top right: active filament; bottom right: arch filament.

prominences. Similarly, 11% of the active objects had $\bar{B}_\parallel \geq 15$ G, while the corresponding percentage was only 4 for quiescent prominences.

Observations indicate that different types of active objects often have magnetic fields of different strength. Table II.6 gives rough values bracketing most of the reported observed fields. Knots may be considered extremely active cap-like prominences, and arch filaments and fibrils may be their disk counterparts. Among these types we find the strongest fields measured in any kind of prominence.

TABLE II.6

Longitudinal component of magnetic field, $B_{\parallel}$, in different types of active prominences

Type	$B_{\parallel}$ (G)	
knots	strong;	100–200
caps	medium-strong;	30–150
loops	medium-strong;	20–100
active filaments	medium;	20– 70
surges		0–150
coronal clouds	} small;	1– 20
streamers		

Loops have medium to strong fields, but at their maximum height they show stronger fields than any other type of prominence at that height (Harvey, 1969). Coronal clouds, which often are found at comparable heights in the corona, have significantly smaller fields. These strong fields in loops may be due to the well-organized fine structure of these prominences, which outline magnetic flux tubes. The changes in strength and polarity of the observed longitudinal component of the magnetic field leave little doubt that the field is directed along the loops.

The field strengths observed in active filaments are consistently greater than the fields threading quiescent prominences, presumably due to the closer association of the former with sunspot regions.

Surges offer particularly interesting data. We would expect that if the surges are being shot out along magnetic lines of force, the observed field strength would depend on the angle between the direction of the surge motion and the line of sight. This is also what is found. When surges move perpendicularly to the line of sight, no field, or a very small field is observed; for others, depending on the tilt of the velocity vector, we see stronger or weaker fields of either polarity. Some active regions produce homologous surges; however, such surges often have magnetic fields of very different strength. This would indicate that these surges either drag their field lines with them, or seriously distort existing field lines, and are not merely being shot out passively along a pre-existing and unchanging field line configuration. Other surges, however, seem to do just that.

Figure II.12 shows the observed, nearly constant magnetic field during the development of a surge, as well as the field in a nearby, but unrelated, coronal cloud. Two hours after the surge had died away the field in the small remaining mound (out of which the surge has been shot) still showed the same field strength and polarity as the surge had showed.

(b) *Disk Observations*

It is often difficult to establish a one-to-one correlation between active prominences seen above the limb with their counterparts observed against the solar disk. In the latter case all look more or less like small filaments. Surges can generally be recognized as such on the disk, while an active filament will look like a fairly long low prominence on the limb. These filaments, which are good indicators of the degree of activity of the active region involved, seem to run along magnetic tubes of force, as do fibrils

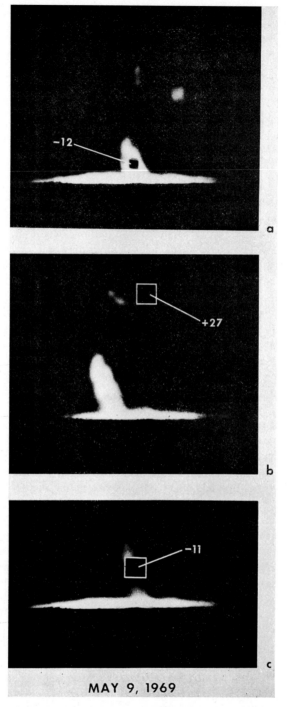

MAY 9, 1969

Fig. II.12. Magnetic field, $B_{\parallel}$, in developing surge prominence, and in nearby coronal cloud. May 9, 1969, a. 17:14 UT, b. 17:39 UT, c. 17:43 UT.

and arch filaments, see the schematic representations in Figure II.11. The fibrils and arches span the neutral line of the longitudinal magnetic field to connect parts of plages of opposite magnetic polarity. The active filaments are nearly always curved, often S-shaped, and even though the greater part of their length is oriented along (above) the neutral line, also these prominences have ends anchored in the magnetic fields of opposite polarities (Malville and Tandberg-Hanssen, 1969; Rust and Smith, 1969).

2.4. Radio, EUV and X-Ray Observations

Nearly everything we know about prominences stems from observations carried out in the optical part of the electromagnetic spectrum. However, information also is contained in other parts of the spectrum, and, in principle, non-electromagnetic means are available. Of the latter, only particle emission is a remote possibility, and has not been explored. On the other hand, some information is emerging from studies of both the long (radio) and the short (EUV and X-rays) wavelength regions of prominence spectra. In the future we can expect such investigations to take on increasingly greater significance.

2.4.1. RADIO OBSERVATIONS

Direct observations of prominences is possible in the mm-wavelength range. Longer waves, in the dm and meter range, are not able to escape from the chromosphere and low corona, and they give information from the outer corona only. This behavior of radio waves is due to the fact that the refractive index n of the solar atmosphere is given by

$$n = \sqrt{1 - \left(\frac{v_c}{v}\right)^2}, \tag{II.49}$$

where v is the frequency of the radio wave, and the critical frequency v_c, the plasma frequency, is

$$v_c = \sqrt{\frac{n_e e^2}{\pi m_e}}. \tag{II.50}$$

Since the refractive index is less than unity, we notice that radio waves are refracted in the opposite sense to light waves when they pass from a tenuous to a dense medium. When $1 - (v_c/v)^2$ becomes negative, propagation is no longer possible. This cut off occurs when the density of the plasma increases so that $v_c = v$. Using a model for the solar atmosphere, we may construct ray paths for radio waves of different frequencies and explore the region of the atmosphere from where the rays can escape, see for instance Pawsey and Bracewell (1955) and Kundu (1965).

The observed temperature at a given frequency and along a ray path is related to the radiation temperature T_R at the optical depth τ_v by the equation

$$T_b = \int_0^{\tau_v} T_R \, e^{-\tau_v} \, d\tau_v. \tag{II.51}$$

The observed temperature is referred to as the brightness temperature, and for an isothermal atmosphere it is simply

$$T_b = T_R(1 - e^{\tau_v}) . \tag{II.52}$$

We see that T_R is an upper limit for the brightness temperature, a limit that is reached for very large optical depth, i.e.,

$$T_b \to T_R \quad \text{as} \quad \tau_v \to \infty . \tag{II.53}$$

For small optical depths, on the other hand, we have

$$T_b \approx \tau_v T_R \qquad \tau_v \ll 1 . \tag{II.54}$$

The optical depth, $d\tau_v = -k_v \, ds$ where ds is a ray path element and k_v the absorption coefficient, is determined mainly by free-free transitions. Generally, the absorption or emission, due to the acceleration, or deceleration, of electrons in the Coulomb field of a charge Ze, is given by

$$\alpha_{ff} = \frac{4\pi Z^2 e^6}{3\sqrt{3} \, hcm^2} \frac{g_{ff}}{v^3 v} , \tag{II.55}$$

where v is the velocity of the free electrons and g_{ff} is the Gaunt factor. If the distribution of velocities is Maxwellian at a temperature T we have

$$k_{ff} = 3 \cdot 69 \times 10^8 \frac{Z^2 n_e n_i}{v^3 T^{1/2}} g_{ff} . \tag{II.56}$$

However, at radio frequencies, where hv/kT is small, the absorption coefficient is (Shklovsky, 1960)

$$k_{ff} = 9 \cdot 8 \times 10^{-3} \frac{Z^2 n_e n_i}{v^2 T^{3/2}} g_{ff} . \tag{II.57}$$

For a fully ionized H plasma with about a 10% He admixture, we may write

$$k_{ff} = \frac{\zeta n_e^2}{v^2 T^{3/2}} , \tag{II.57'}$$

where ζ is a slowly varying function of n_e and T. For chromospheric and prominence condition $\xi \approx 0 \cdot 1$. In the optically thin case we see from Equations (II.54) and (II.57) that

$$T_b \propto \frac{n_e^2}{v^2 T^{1/2}} .$$

Consequently, there exists a limit beyond which further increase in the kinetic temperature of the plasma will correspond to a decrease in the brightness temperature, and therefore to a decrease in the emitted radio flux.

By choosing an appropriate wavelength, which for reasonable models is about $\lambda = 5$ mm, one can explore the atmosphere in and around prominences. Efanov et al. (1968) observed at 8 mm wavelength. Several of their findings are of considerable interest. They compared solar features observed at 8 mm with magnetic maps of the Sun and found that all features of enhanced 8 mm radio emission (what we may call radio plages) are found in regions of increased field strength (magnetic plages). The radio plage lies above the plage as observed in the visible part of the spectrum, and its shape corresponds well with that of the optical plage (see Covington, 1954; Firor, 1959; Kundu, 1959; Christiansen *et al.*, 1960). In such regions one finds quiescent prominences, and their observations of dark filaments indicate a decrease in the level of radio emission relative to the undisturbed Sun, see also Khangildin (1964), i.e., T_b (quiescent prominence) $< T_b$ (quiet Sun), or $\Delta T_b < 0$, where $\Delta T_b = T_b$ (quiescent promi- nence) $- T_b$ (quiet Sun). The brightness temperature for the quiet sun decreases from somewhat more than 7000 K at 8 mm to about 5600 to 5800 K at 1 mm (Reber, 1971).

At slightly longer wavelengths we sample the atmosphere above prominences. It is well known from photovisual coronal observations that there is a dark area above many quiescent prominences, an area of decreased emission. Drago and Felli (1970) observed at $\lambda = 1.95$ cm and found a decrease in T_b over a quiescent prominence. They point out that the decreased radio emission cannot be due to absorption of the photospheric light by the subjacent prominence. Such absorptions must occur in the optical part of the spectrum and even at mm radio waves where the chromosphere is still transparent. But for $\lambda = 1.95$, $\tau_{chrom} \gg 1$. The reason, then, for the radio 'darkening' is due to a lack of coronal emission from layers just above the prominence. This may have a bearing on the process of formation of these objects.

Several studies of the chromosphere at 3 to 4 mm wavelength have included prominence observations.

Buhl and Tlamicha (1970) observed at 3.5 mm and found $\Delta T_b > 0$ in plages and $\Delta T_b < 0$ in dark filaments. In particular, for plages $\Delta T_b \approx 700$ K while several quiescent prominences gave

$$\Delta T_b = -100 \text{ to } -200 \text{ K} .$$

Particularly useful data were obtained during the 1970 eclipse at 3.5 mm by Simon (1971) who found for two filaments (see also Simon and Wickstrøm, 1971),

$$\Delta T_b \approx -400 \text{ K} . \tag{II.58}$$

Few prominence observations are available at $\lambda < 3$ mm. Clark and Park (1968) observed at 1.2 mm with a cooled detector and found fluctuations in the flux over active region. In particular, they found increased emission, $\Delta T_b > 0$, coming from two 'filaments'. From their description it is not obvious what kind of prominences were involved, but they may have been active filaments. The increase in emission amounted to $\Delta T_b \approx 100$ K. At these wavelengths we are already approaching the far IR, where no prominence observations have been made.

There are many observations of the corona at wavelengths longer than a few cm, ranging all the way to meter waves, that may contain information on prominences. When material is shot out from the photosphere during surge activity, radio bursts are generated in the coronal plasma as the prominence on its way out excites radiation at the local electron plasma frequency, ν_e, see Equation (II.50) (Wild and Zirin, 1956; Rosseland and Tandberg-Hanssen, 1957; Riddle, 1970; McCabe and Fischer, 1970; Kundu *et al.*, 1970). However, except for velocity determinations, such observations are difficult to interpret in terms of physical parameters of the prominences. We shall postpone discussion of interaction between prominences and coronal disturbances to Chapter VII.

2.4.2. EUV AND X-RAY OBSERVATIONS

The spectrum below about 3000 Å, called the UV, is heavily absorbed by O_3 in the terrestrial atmosphere, and – below 1500 Å – by O, O_2, and N_2. We shall refer to the spectrum between 1500 Å and 20 Å as the EUV, and to the part below about 20 Å as the X-ray spectrum.

The latter is divided into soft and hard X-rays, and there is no clear-cut definition to help us decide unambiguously where the transition from soft to hard X-rays takes place. This is more a matter of convenience and general practice. We shall designate X-rays of wavelengths longer than about 1 Å as soft. Many people working in this field let the associated energy decide whether the X-rays are soft or hard, calling them hard if the energy exceeds 20 keV (which corresponds to $\lambda < 0.62$ Å).

Rocketborne spectroheliographs have given us monochromatic pictures of the Sun with quiescent prominences in one of the strongest resonance lines in the UV, the K-line of MgII at 2795 Å (Fredga, 1969). The disk filter heliograms taken at 2795 Å are very similar to the CaII, K filter heliograms. However, while quiescent prominences show up fairly distinctly in the CaII, K-line – albeit generally not as clearly as in Hα – the MgII, K filament is faint and fuzzy. The 2795 Å line is a strong line coming from an element more than 10 times as abundant as Ca. However, the MgII line is harder to excite, and probably does not come from layers significantly higher than the CaII line. So far, no quantitative results are available that might further our understanding of prominence conditions beyond what CaII observations have given. Nevertheless, UV measurements are an avenue probably worth exploring further.

In the EUV part of the spectrum we would expect to see strong emission from prominences above the solar limb in the early H Lyman lines, Lyα at 1216 Å and Lyβ at 1026 Å, as well as in the HeI resonance line at 584 Å and in the HeII line at 304 Å. Some of these, as well as other lines, have been observed with instruments onboard two orbiting solar observatories (OSO IV and VI), and preliminary results are available (Noyes, 1972). It is important to note that lines from several times ionized atoms (CIII, OIV, OV, OVI, NeVIII) are seen both in quiescent and in active prominences. The spatial resolution of the satellite-borne equipment is too low to distinguish whether the EUV lines and the Hα line are emitted in nearly the same location. However, in the case of a surge the deceleration in both the Hα plasma and in the EUV plasma was the same (Kirshner and Noyes, 1971), and it seems likely that the EUV region, in this case a CIII line emitting region, is physically connected to the Hα region. This means either that a sufficiently large part of the prominence plasma is at temperatures of the order of 100000 to 200000 K, or that penetration by coronal EUV photons is efficient enough to produce the excitation of the CIII ions. Noyes has shown that the former possibility probably is realized. The multicomponent model deduced from observations in the optical part of the spectrum must therefore be expanded to include these very hot layers in prominences, intermediate in temperature between the cool H, metal and even He region, and the 1 to 2 million degree corona.

Noyes reported that in some cases the line of Hα as well as EUV lines were observed from a single prominence, whereas in another EUV prominence no Hα line could be detected. This would indicate that in the latter case the amount of the Hα emitting plasma was so small that it escaped detection, in the other cases this region may dominate. Therefore, the different types of prominence spectra produced reflect the relative importance in any prominence of the cool, intermediate and hot parts of the plasma present. In this respect also quiescent prominences require a multi-component model, whether or not all lines in the optical part of the spectrum come from the cool component. In other words, all prominences have a multi-layer skin consisting of progressively hotter plasmas as we go toward the outer parts of the prominence, adjacent to the hot corona. The skins or shells, surrounding the cool inner core of all prominence fine-structure threads, are similar to the transition region chromosphere-corona, and pose many of the same problems as the transition region.

When we finally go to the shortest wavelengths of the electromagnetic spectrum, to X-rays of $\lambda < 20$ Å, we find very few pertinent observations. There is a fast-growing literature on X-rays from flares and from active regions in general (see Neupert, 1969; E. v. P. Smith, 1970), but little specifically about prominences.

Vaiana et al. (1968) found that the brightest part of a soft X-ray flare (3 to 14 Å) was located at the position of a quiescent prominence that had undergone a disparition brusque some minutes earlier.

Teske (1971) found soft X-rays between 8 and 12 Å to be emitted from several major loop-prominence systems. It is not unexpected to find that loops contain a high-energy plasma component. We have previously mentioned that loops, which form after major flare events, present us with some of the most energetic aspects of solar activity, and the high-energy plasma fits this picture. Again, we return to the implications that this may have on the theory of loops in Chapters IV and VI.

CHAPTER III

MODELS

Based on data discussed in Chapter II we are now in a position to construct simple models of some types of prominences from deduced values of the following parameters

T_e electron temperature
n_e electron density
ξ_t microturbulent velocity
v macroturbulent velocity
V gross (organized) velocity
L characteristic length
$B_{\parallel}$ longitudinal magnetic field
α filamentary structure ratio (Equation II.21).

We first discuss quiescent prominences for which fairly detailed models are available, and then proceed to the complexities of different types of active prominences.

3.1. Quiescent Prominences

Prior to presenting some of the proposed models, we consider the degree of ionization to be expected in the prominence plasma. This quantity is of crucial importance for the understanding of the physics of prominences, and much work has been devoted to its determination.

3.1.1. DEGREE OF IONIZATION

We have in the two first chapters often alluded to the prominence *plasma*, and from observations of the many lines from ionized metals – Fe, Ti, Ca, etc. – we know that some free electrons are provided by ionization of metals. However, in this connection, the abundance of these metals is so small as to be insignificant compared to H. The question therefore arises as to the role of this latter element in the overall ionization equilibrium.

(a) *Some General Considerations*

Under conditions of local thermodynamic equilibrium (LTE) use of the Boltzmann and Saha equations gives the occupation numbers of different excited and ionized atomic levels in terms of the temperature and density. The source function – given as the ratio at frequency v between the emissivity of the plasma and its absorptivity, $S_v \equiv j_v/k_v$ – is uniquely given by the Planck function at the radiation temperature, T_R,

$$S_v(\text{LTE}) = B_v(T_R) = \frac{2hv^3}{c^2}\left(e^{hv/kT_R} - 1\right)^{-1}.$$

In LTE, the radiation temperature is equal to the electron kinetic temperature. In the radiating atoms and ions, the different energy states will be populated according to Boltzmann's formula

$$\frac{n_U}{n_L} = \frac{g_U}{g_L} e^{\chi_{UL}/kT_{ex}} ,$$

which gives the ratio of the populations of an upper and a lower state in terms of the statistical weights g of the states, the excitation potential χ_{UL} and an excitation temperature parameter, T_{ex}. In LTE, all excitation temperatures are equal to the electron kinetic temperature T_e. The distribution of ions on the different ionization stages is given by Saha's equation

$$\frac{n_{i+1}}{n_i} n_e = \left(\frac{2\pi mkT_{kin}}{k^2}\right)^{3/2} \frac{2U_{i+1}}{U_i} e^{-\chi_i/kT_{ion}} ,$$

where n_i denotes the number density of ions in the ionization stage i, and χ_i is the ionization potential for ionization from stage i to the next stage $i+1$. The ionization temperature, T_{ion}, is a parameter associated with this particular process. In LTE all ionization temperatures, and all kinetic temperatures, T_{kin}, for the different particles are identical to T_e. The partition function for the ionization stage i is given by

$$U_i = \sum_i g_i e^{-\chi_i/kT_{ion}} .$$

Using accepted values for T_e and n_e in quiescent prominences, we could find for instance the fractional ionization of H, $(n_e/n(H))^*$, where $n(H)$ is the total H density, and an asterisk indicates LTE conditions. For H, the degree of ionization is often expressed as $n(H_{II})/n(H_I)$, where $n(H_{II})$ is the proton density.

However, the solar atmosphere, and in particular prominences embedded in this atmosphere, is far removed from LTE conditions. Given this fact, we find that the evaluation of $n_e/n(H)$ is no longer simple, and more sophisticated methods have been tried. As soon as one abandons the simplicity allowed by the assumption of LTE, the specification of the source function becomes of utmost importance. In the presence of spectral lines we may write the source function

$$S_\nu = \frac{j_\kappa + j_1}{k_\kappa + k_1} ,$$

where subscript κ refers to the continuum radiation and subscript l to the radiation in the line. We write

$$S_\nu = \frac{S_1 + r_\nu S_\kappa}{1 + r_\nu} , \tag{III.1}$$

where $r_\nu = k_\kappa/k_1$ is the ratio of the continuum absorption coefficient to the line absorption coefficient at the given frequency. We have the relations

$$d\tau_\nu = -(k_\kappa + k_1) \, ds = d\tau_\kappa + d\tau_1 = (1 + r_\nu) \, d\tau_1 . \tag{III.2}$$

It is convenient to measure the optical depth in terms of the opacity at the line center,

τ_0. If the absorption coefficient is Gaussian over the core of the line, we have (see Equation (II.4)):

$$\tau_v = \tau_0 \, e^{-(\Delta v/\Delta v_D)^2} \, . \tag{III.3}$$

We assume that the source function for the continuum is equal to the Planck function

$$S_\kappa = B(T_e) \, . \tag{III.4}$$

However, the line source function is not equal to the Planck function, but, quite generally, it may be written (Thomas, 1957) as

$$S_l = B_v(T_{ex}) \frac{\psi_v}{\Phi_v} \, , \tag{III.5}$$

where ψ_v is the (normalized) line emission profile and Φ_v the (normalized) line absorption profile. In general, ψ_v/Φ_v varies with frequency across the line. Only when $\psi_v/\Phi_v = 1$ does T_{ex} have its usual meaning of the excitation temperature, and the source function reduces to

$$S_l = B_v(T_{ex}) \, . \tag{III.6}$$

Expression (III.6) is a good approximation for the core of resonance lines. It may be written as

$$S_l = \frac{2hv^3}{c^2} \, (e^{hv/kT_{ex}} - 1)^{-1} = \frac{2hv^3}{c^2} \left(\frac{n_L g_U}{n_U g_L} - 1 \right)^{-1} \tag{III.7}$$

which shows that, to determine the line source function, we must evaluate the ratio of the populations n_L and n_U of the energy levels associated with the line in question. We return to this below.

When the photon scattering processes are non-coherent, and $\psi_v/\Phi_v \neq 1$, the line source function may be written (Warwick, 1955)

$$S_l = \frac{\int J_v \Phi_v \, dv + \varepsilon B}{1 + \varepsilon} \frac{\psi_v}{\Phi_v} \, , \tag{III.8}$$

where the Planck function is assumed constant over the band width of the line. The quantity ε is the photon destruction probability. For a two-level atom it is defined as the ratio of the rates of collisional to spontaneous de-excitation. J_v is the mean specific intensity of radiation, $J_v = \int I_v (d\omega/4\pi)$, and $d\omega$ is an element of solid angle.

To determine the population ratio, n_L/n_U, one assumes LTE conditions and a statistically steady state in each atomic energy level. When collisionally induced transitions are included, this treatment becomes a generalization of the 'principle of reversibility' discussed by Rosseland (1926) and which has been developed by Giovanelli (1948), Thomas (1948a,b), Jefferies (1953), and others.

The equations of statistical equilibrium may be written

$$\frac{dn_j}{dt} = \sum_{\substack{i=1 \\ i \neq j}}^{\kappa} (n_i P_{ij} - n_j P_{ji}) = 0 \, , \qquad J = 1, 2, 3, \dots, \kappa \, . \tag{III.9}$$

P_{ij} is the total rate for transitions from level i to level j and is a sum of two terms,

$P_{ij} = C_{ij} + R_{ij}$, where C_{ij} refers to collisionally induced transitions, while R_{ij} is the radiative transition rate. The solution of Equation (III.9) may be put in the form

$$\frac{n_i}{n_j} = \frac{P^{ij}}{P^{ji}}, \tag{III.10}$$

which dates back to Rosseland (1936) and where P^{ij} is the cofactor of the element P_{ij} in the matrix of the coefficients of Equation (III.9).

To find the source function we would now insert Equation (III.10) into Equation (III.7), but several of the radiative transition rates depend on the radiation field, whence also the ratio n_i/n_j will depend on J_v. The mean intensity, J_v, however, is itself a function of the source function, so we see that the equations involved are all strongly coupled. We need an auxiliary equation, which is the equation of radiative transfer

$$\cos\theta \frac{dI_v}{d\tau_v} = I_v - S_v, \tag{III.11}$$

where $\theta = \cos^{-1}\mu$ is the angle between the path of the photons and the normal to the assumed plane parallel atmosphere. Using Equation (III.2) we may write Equation (III.11) in the form

$$\frac{\mu}{\phi_1 + r_0} \frac{dI_v}{d\tau_0} = I_v - S_v, \tag{III.12}$$

where we have used the relation $r_0 = \phi_v r_1$. The unnormalized absorption profile ϕ_v is related to Φ_v through a constant. In Equation (III.12) S_v is

$$S_v = \frac{r_0 S_\kappa}{\phi_v + r_0} + \frac{\phi_v S_1}{\phi_v + r_0}. \tag{III.12'}$$

The solution of Equation (III.12) has been discussed by a number of authors. In the Eddington approximation it takes the form

$$\frac{1}{3} \frac{d^2 J_v}{d\tau_v^2} = J_v - S_v \tag{III.13}$$

and was solved by Jefferies and Thomas (1958, 1960), by specifying the source function and solving it in terms of a Gaussian quadrature over frequency. A different method to solve Equation (III.12) was derived by Athay and Skumanich (1967) for a line formed in non-coherent, resonance scattering (see Equation (III.8)). They introduced the flux, $H_v = \int \mu I_v (d\omega/4\pi)$, and changed Equations (III.12) to

$$S_v = B_v + \frac{1}{(1+r_0)(\varepsilon+\delta)} \int \frac{\Phi_v}{\phi_v + r_0} \frac{d}{d\tau_0} H_v \, dv. \tag{III.14}$$

Because of the dependence of S_v on $dH_v/d\tau_0$, Equation (III.14) is referred to as the flux derivative form. To derive Equation (III.14) one must still assume equality of Φ_v and ψ_v in the calculation of the flux, but otherwise they need be equal only at the line center, i.e., $\Phi_0/\psi_0 = 1$. The quantity δ introduced in Equation (III.14) is $\delta = r_0 \int (\Phi_v/\phi_v + r_0) \, dv$. Athay and Skumanich solved Equation (III.14) by following an expansion technique for S_v due to Avrett and Loeser (1963). The flux H_v is a well known integral transform of S_v, and the integration can be performed directly. They obtained thereby a matrix operator and regarded the expression for S_v as a matrix equation.

(b) *The Ionization of H in Prominences*

The degree of ionization of H has been evaluated by straightforward methods (Wurm, 1948; Ivanov-Kholodny, 1959; Hirayama, 1963). We shall briefly look at these determinations before we proceed to recent, more comprehensive studies.

Equation (II.19) indicates how the quantity $N_e = n_e X$ can be evaluated from observing the continuous spectrum. Further, the method assumes that $n_e = n(\text{H\,{\sc ii}})$. Therefore, if we know $n(\text{H\,{\sc i}})X$, we can determine the ratio

$$\frac{n_e X}{n(\text{H\,{\sc i}})X} = \frac{n(\text{H\,{\sc ii}})}{n(\text{H\,{\sc i}})} . \qquad (\text{III.15})$$

To determine $n(\text{H\,{\sc i}})X$, one makes the assumption that the degree of ionization of H is equal to the degree of ionization of some singly ionized metal ion with well-observed lines. One of the ions Ti\,{\sc ii}, Sr\,{\sc ii} or Ca\,{\sc ii} is generally used, since their ionization potentials are close to that of H\,{\sc i} (13·6 eV). This leads to equations like

$$\frac{n(\text{M\,{\sc iii}})}{n(\text{M\,{\sc ii}})} = \frac{n(\text{H\,{\sc ii}})}{n(\text{H\,{\sc i}})} ,$$

where M\,{\sc iii} and M\,{\sc ii} stand for doubly and singly ionized metals respectively, for example, Sr\,{\sc iii} and Sr\,{\sc ii}. If one further assumes that neutral as well as more than doubly ionized metal atoms may be neglected compared to $n(\text{M\,{\sc iii}}) + n(\text{M\,{\sc ii}})$, we find

$$\frac{n(\text{H\,{\sc i}})X}{n(\text{M\,{\sc ii}})X} = \frac{n(\text{H\,{\sc ii}})}{n(\text{H\,{\sc iii}})} \frac{n(\text{H\,{\sc i}}) + n(\text{H\,{\sc ii}})}{n(\text{M\,{\sc ii}}) + n(\text{M\,{\sc iii}})} = \frac{n(\text{H})}{n(\text{M})}$$

or

$$n(\text{H\,{\sc i}})X = \frac{n(\text{H})}{n(\text{M})} n(\text{M\,{\sc ii}})X . \qquad (\text{III.16})$$

The quantity $n(\text{M\,{\sc ii}})X$ can be obtained from observing the total energy of selected M\,{\sc ii} lines and from knowing the excitation temperature of the lines involved. Ivanov-Kholodny used Ca\,{\sc ii} lines and found $n(\text{Ca\,{\sc ii}})X = 2·5 \times 10^{13}$ cm^{-2} (see also Vyazanitsyn (1947)). Hirayama used Ti\,{\sc ii} lines, while Wurm analyzed Sr\,{\sc ii} lines.

A knowledge of the abundance of the element in question relative to that of H, $n(\text{M})/n(\text{H})$, will then by Equation (III.16) give us the quantity $n(\text{H\,{\sc i}})X$, which – combined with the relation $n_e = n(\text{H\,{\sc ii}})$ – leads to a value for the degree of ionization, Equation (III.15). The results of the several published analyses do not agree too well within themselves, but it seems that, to a first approximation, the ionization is described by

$$1 < \frac{n(\text{H\,{\sc ii}})}{n(\text{H\,{\sc i}})} < 10 ,$$

or, as fractional ionization $n_e/n(\text{H}) = n(\text{H\,{\sc ii}})/[n(\text{H\,{\sc i}}) + n(\text{H\,{\sc ii}})]$

$$0·5 < \frac{n_e}{n(\text{H})} < 0·9 . \qquad (\text{III.17})$$

If we want to go much beyond the simple calculations above, we find that the study of the excitation and ionization equilibria leads to a complex radiative transfer problem. This question has been tackled by Hirayama (1964), Kawaguchi (1964), Yakovkin and Zel'dina (1964a,b), Giovanelli (1967), and has been solved in more detail by Poland *et al.* (1971). The latter authors used a model atom consisting of two bound levels plus a continuum. They assumed the Lyα line to be saturated, and the Balmer continuum turned out to be optically thin. The Ly continuum transfer problem was solved in detail for a model prominence slab of given thickness (500 km and 6000 km), and with constant total H density, $n(H)$. The inclusion of a density gradient (Poland and Anzer, 1971) introduces only minor changes in the excitation of the H atoms, which may indicate that density fluctuations due to filamentary structures ($\alpha < 1$) will not affect the excitation conditions significantly. The model slab was illuminated from both sides by photospheric ($T_R = 5400$ K) and chromospheric and coronal ($T_R = 6500$ K) radiation fields, and the slab was assumed isothermal.

The electron density, determined by the degree of ionization of H, will vary through such a slab prominence since the radiation field varies through the slab. The fractional ionization, $n_e/n(H)$, is determined by

   (i)   the electron temperature
   (ii) the hydrogen density
   (iii) the impressed radiation fields
   (iv) the thickness of the prominence.

Poland *et al.* (1971) used the flux derivative form (Equation (III.14)) and treated the external radiation field explicitly by separating $J_\nu(\tau_\nu)$ into two parts, $J_\nu(\tau_\nu) = J_\nu^{int}(\tau_\nu) + J_\nu^{ext}(\tau_\nu)$. What we have called the external field, $J_\nu^{ext}(\tau_\nu)$, i.e., that part of the radiation field in the prominence plasma due to radiation from outside, is reduced as it penetrates into the prominence approximately as $e^{-\tau_\nu}$. Therefore, if we let $J_\nu^{imp}(\tau_\nu)$ denote the radiation field which strikes the prominence slab from the outside, we may write

$$J_\nu^{ext}(\tau_\nu) = e^{-\tau_\nu} J_\nu^{imp}(\tau_\nu) .$$

The Lyman continuum portion of the impressed field is known and can be represented by a Planck function at a certain temperature T_R. Poland *et al.* used data due to Zirin *et al.* (1963) and characterized the field by $T_R = 6500$ K. The field is due to coronal emission lines and chromospheric Lyman continuum. The results may be summarized as follows. At low H densities, $n(H) < 10^{11}$ cm^{-3}, the coronal and chromospheric Ly continuum radiation fields penetrate deeply into the prominence. The radiative energy is trapped within the prominence and completely determines the degree of ionization. Under such conditions the ratio $n_e/n(H)$ reaches values in excess of about 0·8 for electron temperatures about 6000 K. At high H densities, $n(H) > 10^{12}$ cm^{-3}, these radiation fields become unimportant in the prominence interior, and the degree of ionization depends on the photospheric radiation field in the Balmer continuum. Under these conditions also the electron temperature plays a stronger role, leading to negligible ionization for $T_e = 6000$ K, while $T_e \geq 7500$ K gives $n_e/n(H) \geq 0·3$ for $n(H) = 10^{12}$ cm^{-3}, see Figure III.1. It is interesting to note that the maximum ionization is not always encountered in the outermost skin of the prominence. This is the case only for a high-density prominence, $n(H) > 10^{12}$ cm^{-3}. For low densities we find that the ioniza-

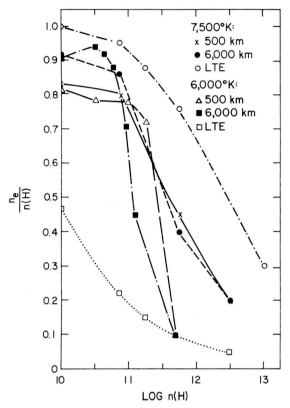

Fig. III.1. Fractional ionization, $n_e/n(H)$, volume average, of an isochoric, isothermal slab, as a function of H density, $n(H)$, for thickness X of 6000 km and 500 km, at temperatures of 6000 K and 7500 K. LTE results are given for comparison.

tion increases as we go into the prominence. This is due to an increase in the density of the radiation field with depth caused by trapping of the field. For quiescent prominences whose density is about $n(H) \approx 10^{11}$ cm^{-3} or less, the degree of ionization seems to be given by

$$0 \cdot 7 < \frac{n_e}{n(H)} < 0 \cdot 9 \,. \tag{III.18}$$

A comparison of expressions (III.17) and (III.18) indicates that the smaller values sometimes quoted for $n_e/n(H)$ probably underestimate the degree of ionization of prominence material.

3.1.2. Models without Magnetic Field

The models that attempt to describe quiescent prominences generally have considered such objects to be in a quasi-equilibrium state (except during periods of activation). Therefore, if magnetic fields are neglected, pressure equilibrium must reign at any given height, i.e.,

$$n_{\text{prom}} k T_{\text{prom}} \approx n_{\text{cor}} k T_{\text{cor}} \,. \tag{III.19}$$

At heights of 25000 to 50000 km, where the electron density of the ambient corona may be about 2 to 4×10^8 cm^{-3}, and where the temperature is 1·5 to $2 \cdot 0 \times 10^6$ K (Newkirk, 1967), we find that the density in a 5000 K prominence should be $n_{\mathrm{prom}} \approx 10^{11}$ to 5×10^{11} cm^{-3}. If one finds values of n_{prom} significantly different from the above range, this implies that the prominence is in some sort of dynamic equilibrium, and the static models fail, or that the magnetic pressure must be taken into account.

TABLE III.1

Models of quiescent prominences, $\mathbf{B}=0$

Quantity	Author			
	de Jager	Ivanov-Kholodny	Hirayama	Jefferies and Orrall
T_e (K)	$1 \cdot 5 \times 10^4$	6×10^3	$4 \cdot 5 \times 10^3$	$5 \cdot 3 \times 10^3 - 1 \cdot 0 \times 10^4$ H lines
		9×10^3	8×10^3	$1 \cdot 0 \times 10^4 - 1 \cdot 2 \times 10^4$ H continuum $1 \cdot 5 \times 10^4$ He
T_{ex} (K)	4×10^3 $\chi \approx 3$ eV,	—	$4 \cdot 2 \times 10^3$ Hα	
	$1 \cdot 5 \times 10^4$ $\chi \approx 10$ eV		5×10^3 He I triplets	
n_e (cm^{-2})	2×10^{10}	10^{12}	10^{11}	$8 \cdot 1 \times 10^9 - 5 \times 10^{10}$
V_t (km s^{-1})	4–10	6	6	—
α	1·0	<0·1	0·3	—

Table III.1 summarizes the models of quiescent prominences published by de Jager (1959), Ivanov-Kholodny (1959), Hirayama (1963, 1964) and Jefferies and Orrall (1963). Notice that most models published so far have assumed LTE conditions.

The excellent review article by de Jager is a valuable source of information. However, the kinetic temperature given there is too high. de Jager used data on Hα line profiles that were corrected for self-absorption (Conway, 1952) and that were compared with profiles of lines from heavier elements to determine the effect of microturbulence (Equations (II.8) and (II.9)). As mentioned in Chapter II this is a difficult procedure and may, at times, not be applicable.

The density quoted by Ivanov-Kholodny is much too high, and would lead to a considerable pressure – imbalance (Equation (III.19)). The density was found from a Stark analysis of the wings of prominence lines – a precarious procedure.

Both Hirayama's, and Jefferies and Orrall's models seem to portray temperature and density conditions in quiescent prominences well. The latter also allows for the possibility of a multicomponent model, thereby being able to account for the emission from ionized He – and other high-excitation lines.

However, it should be pointed out that the models without a magnetic field must be incorrect. The pressure balance (Equation (III.19)) shows that the prominence is denser than its surroundings, and it will fall under gravity. A magnetic field can provide the necessary support through the Lorentz force.

3.1.3. Models including magnetic fields

(a) *Some General Comments on Magnetohydrostatic Models*

When a magnetic field is present in the prominence plasma, Equation (III.19) may no longer be valid. However, it cannot simply be modified by adding magnetic pressure terms, $\nabla(B^2/8\pi)$, to gas pressure gradients, ∇p. This would assume that the fields had no twist, and in actual cases the tension, $(\mathbf{B}_{prom} \cdot \nabla)\,\mathbf{B}_{prom}$ for instance, resulting from a twist, may cancel part of the $\nabla(B_{prom}^2/8\pi)$ term.

The structure of magnetic fields in prominences raises several interesting problems. In a steady state the equation of motion reduces to

$$-\nabla p + \frac{1}{c}\mathbf{j}\times\mathbf{B} + \rho\mathbf{g} = 0\,. \tag{III.20}$$

Equation (III.20) shows that when pressure equilibrium reigns, the gravitational pull can be balanced by a Lorentz force. In other words, for certain magnetic field configurations one might expect the prominence material to be supported against gravity by the action of the magnetic field.

An order of magnitude analysis of the terms in Equation (III.20) shows that for reasonable values of the parameters, the $(1/c)\,\mathbf{j}\times\mathbf{B}$ term will dominate. Consequently, we conclude that in quiescent prominences a large part of the current $\mathbf{j}$ is parallel to the field $\mathbf{B}$, or, stated differently, a large part of the prominence field must be force free.

When the electric conductivity is sufficiently high, the condition that a field is force free can be written as

$$\nabla\times\mathbf{B} = \alpha(\mathbf{r},t)\mathbf{B}\,, \tag{III.21}$$

where we have indicated that in the general case the quantity α may be a function of both space and time. In a steady state this scalar function of space is called the reciprocal pitch of the field, and is not to be confused with the filamentary structure ratio (Equation (II.24′). Equation (III.21)) and the condition

$$\nabla\cdot\mathbf{B} = 0 \tag{III.22}$$

define the force-free field. Let us first consider the case $\alpha =$ constant. Then general solutions of Equations (III.21) and (III.22) are possible (Chandrasekhar and Kendall, 1957), and for simple cylindrical symmetry (Lüst and Schlüter, 1954; Schlüter, 1957a) the solution is relatively straightforward (see also Schatzman (1961)). However, there is no reason to assume that α should be constant for magnetic fields in and around prominences. For the case of non-constant α no general method for solving the equations is known, even though solutions have been found for some simple geometrical situations (Ferraro and Plumpton, 1966). See also papers by Grad and Rubin (1958), Gold (1964), Molodensky (1966), Schmidt (1966, 1968), Sturrock and Woodbury (1967), Jette and Sreenivasan (1969), Nakagawa *et al.* (1971), and Raadu and Nakagawa (1971).

Another question concerning the field configuration is the orientation of the field with respect to the long axis of a quiescent prominence. Most early models assumed the field to be nearly perpendicular to the long axis, and we shall treat these in Section b. However, there has been growing evidence that a part, maybe most, of the magnetic flux is channeled along the prominence, and this case will be treated in Section c. In Section d we shall consider the fascinating questions related to twisted magnetic fields.

So far one has been able to measure only the longitudinal component, $B_{\parallel}$, of the magnetic field, and only rarely does this component lie along or transverse to the prominence axis. In the following discussion we shall consider an orthogonal coordinate system with x-axis horizontal and perpendicular to the long axis of the prominence, y-axis horizontal and along the same axis, and z-axis vertical. The observations of a prominence on the solar limb near the equator and in the plane of the sky give $B_{\parallel} = B_x$, see Figure III.2, prominence 1. In the general case, for a prominence at latitude ϕ and not in the plane of the sky, to find B_x one must – near the central meridian – measure the angle θ that the assumed vertical plane filament makes with the north-south direction on the Sun, see Figure III.2, prominence 2. We return to this in Section c.

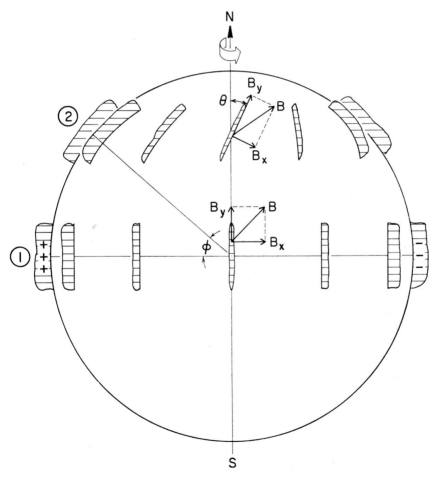

Fig. III.2. Parameters defining (a) a prominence in the plane of the sky on the solar equator (Prominence 1) and (b) a prominence at latitude ϕ and not in the plane of the sky (Prominence 2).

(b) *Field Configurations Capable of Supporting Prominences*

The basic idea in these models is that the prominence material is supported against gravity by the Lorentz force, $\mathbf{j} \times \mathbf{B}$, where the field lines lie in planes perpendicular to the prominence sheet, and the current flows along the prominence (the y-direction). The models describe the prominence material in mechanical equilibrium under the combined actions of gas pressure, gravity and the Lorentz force, other forces being neglected. Equation (III.20) with $\mathbf{j} = (1/4\pi)\nabla \times \mathbf{B}$ then reads

$$\nabla p - \rho \mathbf{g} - \frac{1}{4\pi}(\nabla \times \mathbf{B}) \times \mathbf{B} = 0 \,. \tag{III.23}$$

These three equations, together with Equation (III.22) and an equation describing the atmospheric model, will give us 5 equations for the five unknowns, $\mathbf{B}$, p and ρ. Menzel was the first to try to show that coronal magnetic fields can support quiescent prominences in a static equilibrium model (Bhatnagar *et al.*, 1951). The magnetic field lines were supposed to be in planes parallel to the long axis of the prominence. We shall apply Menzel's analysis to a field perpendicular to the long axis. Somewhat different formulations of the problem are due to Dungey (1953) and Kippenhahn and Schlüter (1957). A particularly lucid treatment by Brown (1958) which shows the interrelation between the different models, will partly be followed here.

In our coordinate system all these models assume that $\mathbf{B}_y = 0$ and that p, ρ and $\mathbf{B}$ are independent of y. Furthermore, the atmosphere is isothermal so that the last equation we need reads

$$p = nkT = \rho g H_0 \,, \tag{III.24}$$

where $H_0 = kT/mg$ is the scale height in the absence of a magnetic field. Let F be a scalar function, $F = F(x, z)$ then from Equation (III.22), we may state

$$B_x = -\frac{\partial F}{\partial z}, \quad B_z = \frac{\partial F}{\partial x} \,. \tag{III.25}$$

This means that Equation (III.23) may be written

$$\nabla p - \rho g = -\frac{1}{4\pi}(\nabla^2 F)\nabla F \,. \tag{III.26}$$

If we combine Equations (III.24) and (III.26) we can write

$$\nabla[p \, e^{z/H_0}] = -\frac{1}{4\pi}[\nabla^2 F \, e^{z/H_0}]\nabla F \,.$$

This shows that $p \, e^{z/H_0}$ is a function of F, $p(F)$ say, and that

$$\frac{\mathrm{d}}{\mathrm{d}F}[p(F)] = -\frac{1}{4\pi}\nabla^2 F \, e^{z/H_0} \,.$$

Then $\nabla^2 F \, e^{z/H_0}$ must be a function of F and the scalar function satisfies the equation

$$\nabla^2 F = \phi(F) \, e^{-z/H_0} \,, \tag{III.27}$$

where $\phi(F)$ is any arbitrary function of F.

Brown (1958) considered Equation (III.27) the basic equation for static equilibrium in this two-dimensional case. Any function F that is a solution of Equation (III.27) will give a possible model for the magnetic field, Equation (III.25). The ensuing pressure distribution is then given by

$$p = -\frac{1}{4\pi} e^{-z/H_0} \int \phi(F) \, dF .$$

TABLE III.2

Models of magnetic field in quiescent prominences, $B_y = 0$

Author	$\phi(F)$	F
Menzel (modified)	$\phi(F) = AF^{(1-2H/H_0)}$ $A = $ const.	$F = F_1(x) e^{-z/2H}$ $\dfrac{d^2 F_1}{dx^2} + \dfrac{F_1}{4H^2} = AF_1^{(1-2H/H_0)}$
Dungey	$\phi(F) = D = $ const.	$F = DH_0^2 e^{-a/H_0} + F_2$ $F_2 = 1 - e^{-z/H_0} - 2 e^{-z/2H_0} \cos \dfrac{x}{2H_0}$
Kippenhahn and Schlüter	$\phi(F) = C e^{F/GH_0}$ $C = $ const. < 0 $G = $ const.	$F = Gz + F_3(x)$

Solutions due to Menzel, Dungey (1953), and Kippenhahn and Schlüter (1957) are shown in Table III.2, in terms of the choice of function $\phi(F)$ and the form of the scalar function F.

In the framework of Equation (III.27) Menzel's model corresponds to the choice

$$\phi(F) = AF^{(1-2H/H_0)} , \tag{III.28}$$

where A is a constant. In the original paper, Menzel remarked that Equation (III.26) is separable if the function $p(F)$ has the form $p_1(x) e^{-z/H}$ and if F can be written as $F = F_1(x) e^{-z/2H}$, which is equivalent to the choice (III.28) and leads to the following differential equation for F_1,

$$\frac{d^2 F_1}{dx^2} + \frac{F_1}{4H^2} = AF_1^{(1-2H/H_0)} .$$

This equation integrates to

$$\left(\frac{dF_1}{dx}\right)^2 + \frac{F_1^2}{4H^2} = c - \frac{A}{qF_1^{2q}} ,$$

where c is a constant, and where $H/H_0 = q + 1 > 1$. If we assume that the lines of force are horizontal as they traverse the prominence sheet, i.e., $dF_1/dx |_{x=0} = 0$, we find

$$\left(\frac{dF_1}{dx}\right)^2 + \frac{F_1^2}{4H^2} = \frac{F_1^2(0)}{4H^2} + \frac{A}{q} [F_1(0)^{-2q} - F_1^{-2a}] ,$$

and the corresponding pressure is

$$p = \frac{A}{8\pi q} F^{-2q} e^{-z/H} .$$

In terms of the ratio γ between gas and magnetic pressure, $\gamma = 4H^2 A/(q F_1(0)^{2q+2})$, the equation for F_1 takes the form

$$4H^2 \left(\frac{dF_1}{dx}\right)^2 = F_1(0)^2(1+\gamma) - F_1^2 - \gamma F_1(0)^2 \left(\frac{F_1(0)^{2q}}{F_1}\right) \tag{III.29}$$

and the pressure can be written as

$$p = \frac{\gamma F_1(0)^2}{32\pi H^2} \left(\frac{F_1(0)}{F_1}\right)^{2q} e^{-z/H} . \tag{III.30}$$

Equation (III.29) must be solved numerically, but some general physical properties may be inferred without solving it. The lines of force, along which $F =$ constant, form a set of curves that may be described by the equation

$$F_1 = e^{(z-z_0)/2H} ,$$

where z_0 is a constant that varies from line to line. For $\gamma q > 1$ F_1 has a minimum at $x = 0$, which means that the lines of force are bowed to a minimum height in the prominence sheet. Furthermore, the pressure has a maximum at $x = 0$, and the model consequently gives a rough picture of how prominence material is denser than the surroundings and causes a slight sag of the lines of force in the prominence (see Figure III.3).

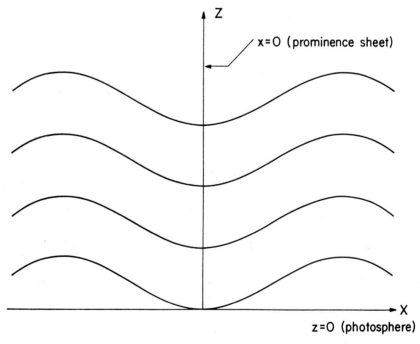

Fig. III.3. The supporting magnetic field in Menzel's model.

An obviously simple choice for the function $\phi(F)$ is a constant, $\phi(F)=D$. Then

$$F = DH_0^2 e^{-z/H_0} + F_2$$

is a solution for F. Dungey gave his solution for

$$F_2 = 1 + e^{-z/H_0} - 2 e^{-z/2H_0} \cos \frac{x}{2H_0} . \tag{III.31}$$

In this case the lines of force are closed loops for $F < 1$ and infinite wavy lines for $F > 1$. The magnetic field is assumed to be zero except between two closed loops, $F=a$ and $F=b$, so that the pressure outside the outer loop $F=a$ is not affected by the field. Inside the loop $F=b$, the pressure is greater than outside $F=a$ at the same height. By redistributing the matter inside $F=a$ (the prominence), the magnetic field may ensure that a narrow horizontal filament floats in equilibrium, surrounded by material of considerably smaller density.

Cowling (1957) criticized this model and drew attention to the fact that Dungey's solution requires currents running in opposite directions near $F=a$ and $F=b$, which hardly seems a likely situation. However, by letting the prominence exist inside closed loops of magnetic field, Dungey introduced a field configuration that may be of importance in the study of prominence stability, and may have a bearing on the question of twisted magnetic fields. We shall return to these aspects later.

The third model, due to Kippenhahn and Schlüter, corresponds to the choice, see Table III.2.

$$\phi(F) = C e^{F/GH_0} \tag{III.32}$$

in Equation (III.27), where C and G are constants and the trial solution has the form

$$F = Gz + F_3(x) .$$

We obtain an equation for $F_3(x)$ that can be integrated directly.

In their original paper, Kippenhahn and Schlüter (1957) combined Equations (III.23), (III.24) and (III.25) to derive the following equation for the field

$$B_x \nabla^2 B_z - B_z \nabla^2 B_x + \frac{B_z}{H_0} \left[\frac{\partial B_z}{\partial x} - \frac{\partial B_x}{\partial z} \right] = 0 . \tag{III.33}$$

The density distribution $\rho(x,z)$ follows from the x-component of Equation (III.23):

$$\frac{\partial \rho}{\partial x} = -\frac{mB_z}{4\pi kT} \left[\frac{\partial B_z}{\partial x} - \frac{\partial B_x}{\partial z} \right] . \tag{III.34}$$

The authors argued that B_x will vary slowly with x in a thin prominence, and took B_x independent of x. They further assumed B_x to be independent also of height, in other words $B_x = \text{const}$. By Equation (III.22) B_z will not depend on z. These conditions simplify Equation (III.33) to a differential equation for B_z

$$\frac{\partial^2 B_z}{\partial x^2} + \frac{B_z}{H_0 B_x} \frac{\partial B_z}{\partial x} = 0 . \tag{III.35}$$

The boundary conditions $B_z(x=0)=0$ and $B_z(x \to \infty) = B_z(\infty)$ lead to the following solution of Equation (III.35),

$$B_z = B_z(\infty) \tanh \left[\frac{B_z(\infty)}{B_x} \frac{x}{2H} \right], \qquad\qquad (\text{III.36})$$

where $B_z(\infty)/B_x > 0$. The corresponding density distribution, Equation (III.34), becomes

$$\frac{\partial \rho}{\partial x} = -\frac{m}{8\pi kT} \frac{\partial B_z^2}{\partial x}$$

and yields the solution

$$\rho = -\frac{mB_z(\infty)^2}{8\pi kT} \left[\tanh^2 \left(\frac{B_z(\infty)}{B_x} \frac{x}{2H} \right) - 1 \right] \qquad\qquad (\text{III.37})$$

for the boundary condition $\rho(x \to \infty) = 0$. Equation (III.36) indicates that the field lines bend down as they traverse the filament, in a manner not unlike the central part of the lines in Menzel's model. The density distribution, Equation (III.37), shows a fairly sharp maximum in the prominence sheet, ($x = 0$), and falls to half its maximum value at $x = \pm 1 \cdot 8 H_0 [B_x/B_z(\infty)]$. The maximum value is $\rho_c = mB_z(\infty)^2/8\pi kT$, which, for a 5000 K prominence whose supporting field lines correspond to $B_z(\infty)$ equal to 2 or 3 G, amounts to about 5×10^{-13} g cm^{-3}, or a H density of $n(\text{H}) \approx 3 \times 10^{11}$ cm^{-3}.

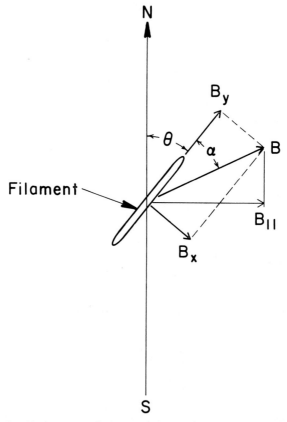

Fig. III.4. The relationship between the angles α and θ, and the components of the magnetic field vector **B**.

It would be of interest to systematically examine a variety of $\phi(F)$ functions, using Brown's treatment. The analysis first should be extended to the more realistic case of a non-isothermal atmosphere. By an appropriate computer program one then could search for more realistic field configurations, subject to observational restraints.

(c) *The Importance of a B_y Component*

For a number of years the Kippenhahn-Schlüter model provided a valuable frame of reference in which to discuss prominence support and stability, and it still has many valuable characteristics. However, doubt as to the adequacy of this particular choice of a two-dimensional approach became evident with Rust's (1966) work on the field configuration. Since magnetographs have not been able to furnish direct data on the strength of the component of the magnetic field along the prominence, B_y, relative to the component perpendicular to its long axis, B_x, statistical methods are used (Harvey, 1969; Tandberg-Hanssen and Anzer, 1970). The latter authors determined the angle θ (see Figure III.2) for a number of prominences whose longitudinal field $B_{||}$ had been measured. From such data one can calculate a mean value $\bar{B}_{||}$ for each value of θ and in principle one can describe the relationship by a function, $F_{obt}(\theta)$. However, of prime interest is the angle α between the filament axis and the magnetic field (see Figure III.4). Tandberg-Hanssen and Anzer assumed a horizontal field, although a z-component may be added without changing the results, and related the observed field to the actual field by

$$B_{||} = B_0|\sin(\alpha+\theta)| \, . \tag{III.38}$$

They further assumed that the horizontal magnetic field vectors in quiescent prominences can be described by a distribution function

$$f(B_0, \alpha)\,\mathrm{d}B_0\,\mathrm{d}\alpha = \frac{N}{N_{tot}}\,\mathrm{d}B_0\,\mathrm{d}\alpha \, , \tag{III.39}$$

where N is the number of filaments having a field of strength B_0 making an angle α with the y-axis and N_{tot} is the total number of prominences studied. Equations (III.38) and (III.39) give the average line-of-sight field

$$\bar{B}_{||}(\theta) = \int_0^\infty \int_0^\pi B_0|\sin(\alpha+\theta)|f(B_0, \alpha)\,\mathrm{d}\alpha\,\mathrm{d}B_0 \, . \tag{III.40}$$

The further assumption was made that the distribution function is separable, i.e., $f(B_0, \alpha)=f_1(B_0)f_2(\alpha)$, which holds if there is no correlation between the field strength and the angle α. From observations they found $F(\theta)=\overline{B_{||}(\theta)}/\bar{B}_0$, and $f_2(\alpha)$ was determined such that

$$F(\theta) = \int_0^{\pi/2} |\sin(\alpha+\theta)|f_2(\alpha)\,\mathrm{d}\alpha \, .$$

Notice that the case of $f_2(\alpha)=\frac{1}{2}[\delta(\alpha-\alpha_0)+\delta(\alpha-(\pi-\alpha_0))]$ will, for $\alpha_0=\pi/2$, give a field parallel to the x-axis, i.e, as in the Kippenhahn-Schlüter configuration. However, this solution leads to a function $F(\theta)$ quite different from the observed one. Of the solutions for trial angular distribution functions

$$f_2(\alpha) = \frac{2n+1}{2} \left(\frac{2}{\pi}\right)^{2n+1} \left(\frac{\pi}{2} - \alpha\right)^{2n}$$

the one that best fits the data is for $n=2$, which corresponds to a mean value $\bar{\alpha} \approx 15°$. Consequently, for the sample studied by Tandberg-Hanssen and Anzer it seems safe to state that the magnetic field traverses quiescent prominences under a small angle α, or that

$$B_y > 2B_x .$$ (III.41)

(The actual values $\bar{\alpha} \approx 15°$ corresponds to $B_y \approx 4B_x$.)

(d) Helical Field Configurations

Under good seeing conditions $H\alpha$ pictures of some prominences seem to reveal sign of a helical structure. The apparent presence of helical structure often becomes more pronounced during the disparition brusque phase (see Figure II.2) but such a configuration may have characterized also the stable phase of the prominence. Malville (1969) has studied extensively quiescent prominences in search of helical structure, and concludes from his observations that this is not an uncommon characteristic. Since the electric conductivity of the prominence plasma is very high, it seems natural to assume that the fine structure of material as outlined in $H\alpha$ pictures is the same as the fine structure of the magnetic field. We are thus led to study the implications of helical magnetic fields in prominences, see also Rompolt (1971) and Øhman (1972).

On the other hand, quiescent prominences owe their support to a field that is anchored in the photosphere on both sides of the filament. Consequently, we seem to be dealing with two magnetic field components: one, the previously discussed supporting field, which we designate B_0, and another, which we shall call the internal field and designate B_1. The field $\mathbf{B_0}$ is produced by photospheric or subphotospheric currents as assumed previously. For the latter field two cases have been considered. Ioshpa (1968) studied a $\mathbf{B_1}$ field directed along the prominence (i.e, along the y-axis) and bent down into the photosphere at the ends of the filament. Anzer and Tandberg-Hanssen (1970) considered the internal field due to currents flowing along the prominence. As we shall see below, this may result in helical field configurations. Also Nakagawa and Malville (1969) have considered a $\mathbf{B_1}$ field along the prominence, and used it to study the stability of prominences (see Chapter V).

Ioshpa argued that if the internal field is attached to the photosphere at the ends of the prominence, this will lead to great stability of the filament. He estimated the strength of the field necessary to produce stability by comparing the tension of the magnetic lines of force arising from the bending of the filament, $B_1^2/4\pi R_c$, with the gravitational force responsible for the bending, ρg. R_c is the radius of curvature of the lines of force, and $R_c \approx \lambda/4$ for small perturbations, where λ measures the scale of the perturbation [$\sin(2\pi y/\lambda)$]. Ioshpa took λ equal to the thickness X of the prominence $\approx 10^9$ cm, and $n=3 \times 10^{11}$ cm^{-3}, to arrive at $B_1 \approx 10$ G. This value seems reasonable, but it is not obvious where the currents, responsible for the B_1 field, flow in Ioshpa's model, nor what the effect will be of the actual superposition of the two fields in the prominence.

If the current flows along the filament, assumed to be infinitely long, $j_1 = (0, j_y, 0)$, and the internal field will be given by (Anzer and Tandberg-Hanssen, 1970)

$$B_1 = -\frac{1}{2}j_0 \begin{cases} (z, 0, -x) & \text{for } r \leq R, \\ \frac{R^2}{r^2}z, 0, -\frac{R^2}{r^2}x & \text{for } r > R, \end{cases}$$

where

$$j_y = \begin{cases} j_0 & \text{for } r \leq R \\ 0 & \text{for } r > R \end{cases}$$

and $j_0 = $ const. R is the radius of the circularly cylindrical model prominence and r is the distance from its axis. The supporting field is assumed to be initially horizontal, $\mathbf{B}_0 = (B_{0,x}, B_{0,y}, 0)$. Anzer and Tandberg-Hanssen measured $\mathbf{B}_1$ in units of $B_{0,x}$, and defined the dimensionless parameter C, giving the ratio of the magnetic field produced at $r = R$ by the current j_y to the field component $B_{0,x}$ existing in the absence of a prominence,

$$C \equiv \frac{j_0 R}{2B_{0,x}} = \frac{J}{2\pi R B_{0,x}}, \tag{III.42}$$

where $J = \pi R^2 j_0$ is the total current in the prominence.

The total field is given by the superposition of the fields $\mathbf{B}_0$ and $\mathbf{B}_1$, i.e. $\mathbf{B} = \mathbf{B}_0 + \mathbf{B}_1$. Let us study the shape of the field lines projected on planes $y = $ const. The field lines are calculated from the flux function of the field, $F(x, z)$. We have

$$F(x, z) = \int_L \mathbf{B}_0 \cdot \mathbf{n} \, dl, \tag{III.43}$$

where the integration is along a path L and where $\mathbf{n}$ is the normal to the line element of L. The curves $F = $ const. then represent the field lines. The components of the field vector are given by Equations (III.25)., i.e,

$$B_x = \frac{\partial F}{\partial z}, \quad B_z = -\frac{\partial F}{\partial x}.$$

When the supporting field dominates the internal field, i.e., when C is small, the field lines are open (see Figure III.5 where the density of lines indicates the strength of the field). As the relative importance of the internal field increases (C increases beyond unity), a larger and larger part of the prominence is filled with closed field lines. Since B_0 has a y-component, these closed field lines represent in reality helices along the long axis of the prominence.

Anzer and Tandberg-Hanssen made some numerical estimates for a quiescent prominence with $B_{0,x} = 5$ to 10 G, $n_e = 10^{10} - 5 \times 10^{10}$ cm^{-3} and $R = 30\,000$ to $60\,000$ km, considered to be in equilibrium such that the Lorentz force per unit length of filament, $(1/4\pi)JB_{0,x} = \frac{1}{2}CRB_{0,x}^2$, balances the gravitational force, $\pi R^2 \rho g$. The equilibrium condition gives

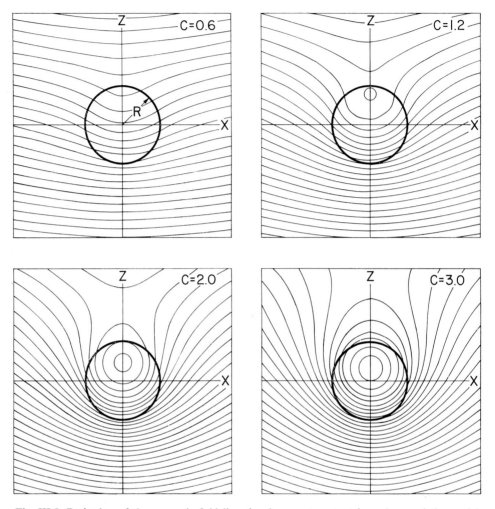

Fig. III.5. Projection of the magnetic field lines in planes $y=$ constant in and around the model
prominence (heavy circle) for different values of the parameter C.

$$C = \frac{\frac{1}{4}R\rho g}{(1/8\pi)B_{0,x}^2}$$
(III.42′)

which states that the parameter C is also the ratio between the potential energy of
prominence material lifted to a height of $\frac{1}{4}R$, and the energy associated with the
magnetic field $B_{0,x}$ in the prominence. For the values of the physical parameters listed
above, C lies in the range $0{\cdot}1 < C < 15$. Such values of C, and the high electric conductivity of the plasma, indicate that in prominences the material distribution and the
magnetic field are strongly coupled.

A value of $C=3$, which leads to a helical magnetic field in most of the prominence
(see Figure III.5) can be realized with the following values of the parameters:

$$B_{0,x} = 8\ \text{G}, \quad n = 1{\cdot}4 \times 10^{11}\ \text{cm}^{-3}, \quad R = 50000\ \text{km}.$$

The cylindrical shape of the prominence model is not realistic, except perhaps during the disparition brusque phase. We will return to this question in Chapter V.

3.2. Active Prominences

3.2.1. INTRODUCTORY REMARKS

Active prominences must be described by more complicated dynamic models than quiescent objects, and for this reason no model has been worked out in any detail. However, when we restrict ourselves to excitation and ionization conditions, we are on somewhat safer ground. The frequent presence in the spectra of very broad lines of He II strongly suggests a multicomponent model (see Section 2.1.3). Reasonable confidence can be put in density and temperature determinations for the cooler regions of the multicomponent models. We find that in this respect these regions resemble the plasma in quiescent prominences, even though the macroscopic velocities are generally much greater.

It is more difficult to ascertain the conditions in the hotter regions, whence comes for instance the He II radiation. It is probable that these regions in loops, that form after large flares, present us with the hottest prominence plasma found in the solar atmosphere and observable in the optical part of the spectrum. Similar comments hold for the plasma responsible for the EUV lines.

We have seen in Chapter II that many of these active prominences are pervaded by magnetic fields, and this coupled with the fact that often high velocities are observed – imposes stringent conditions on the physical models. Material must continuously be fed into the prominence from the surrounding corona or from the chromosphere and photosphere below. We shall refer to the former case as condensation, and various aspects of the physics of this process have been discussed in the literature. The problem of the formation of prominences will be treated in Chapter IV, here we mention merely some aspects of it related to the models.

The case where material is replenished from below is of prime importance in spicules as well as in surges, sprays, and maybe in other types of prominences. Such objects may resemble explosions and therefore reveal a very different type of phenomenon than condensation. We shall refer to this type of formation as injection, as the word explosion may convey the picture of too violent and unorganized a process when we think of the well-organized motions observed in repeating surges.

For quiescent prominences, once the object has been formed, to a first approximation, the model should describe the nearly static conditions. In active prominences, the dynamics is important, since formation and disappearance of parts of prominences go on continuously at a much greater rate. Consequently, it is difficult to separate the concept of formation and model in these cases.

3.2.2. EXCITATION AND IONIZATION OF HELIUM

The radiation from prominences due to He presents us with several important problems. In spectra of quiescent prominences the lines of He II are weak, but they show up with considerable intensity in active objects; at times they exceed the intensity of He I lines. This points to a high degree of excitation of the emitting plasma.

(a) *Neutral Helium*

Lines of orthohelium (the triplet series) as well as lines of parahelium (from the singlet system) are present in all prominence spectra. Because of the selection rules for radiation ($\Delta S = 0$) the two systems behave almost as two separate atoms, except when collisions are important. Referring to Figure III.6 we note that the ground state 1^1S lies more than 20 eV below the next lowest singlet level. The metastable 2^3S level acts as a sort of ground level for the triplet series. This metastable level can be populated in two ways, either by photoionization followed by cascade to the level, or by collisions from the ground state.

In the cooler regions of quiescent prominences it is likely that the former mechanism dominates, and it also may play a role in the low-excitation plasma of some active

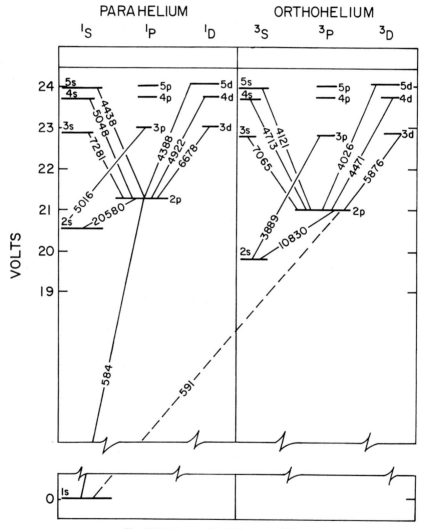

Fig. III.6. Energy-level diagram of He I.

objects. The idea that a relatively high degree of ionization of He could be due to UV radiation from the corona ($\lambda < 504$ Å) was first realized by Goldberg (1939), and was assumed to apply in the chromosphere (see also Miyamoto (1947)). Hirayama (1963, 1964, 1971a) has adopted this point of view for the He emission from quiescent prominences, and extensive work on both He I and He II emission has been done by Yakovkin and Zel'dina (1969, 1971) (see also Burns (1970)).

When we consider active objects, the problem is more involved, since now the density may be high enough for collisions to compete. If the 2^3S level is populated by collisions, it will lead to a pseudo-Boltzmann equilibrium between 2^3S and the ground state,

$$\frac{n(2^3S)}{n(1^1S)} = \frac{g(2^3S)}{g(1^1S)} e^{-\chi(2^3S)/kT_{ex}},$$ (III.44)

with

$$\frac{g(2^3S)}{g(1^1S)} = 3.$$

Also the 2^1S state is populated by roughly the same amount of collision, but this level can be easily depopulated through the 2^1P level, from whence allowed transitions lead to the ground state, giving rise to emission in the resonance line at 584 Å. This goes so fast that the 2^1P level will be underpopulated relative to the 2^3P level. Consequently, we expect the ratio between lines whose lower level is 2^3P and lines whose lower level is 2^1P to be larger than the value corresponding to LTE, which is 3:1. This is exactly what is observed in all quiescent prominences (see Table III.3) which pertains to observations made at the HAO Climax station, and which indicates that the triplet/singlet line intensity ratio is more than 10 times the LTE value in these

TABLE III.3

The triplet/singlet line intensity ratios, $I_3/I_1 \equiv I(n^3D - 2^3P)/I(n^1D - 2^1P)$, in some active (A) and quiescent (Q) prominences

Date	Type of prominence	n	Wavelengths of lines	I_3/I_1	Mean value of I_3/I_1
June 14, 1956	Q	3	$\frac{5876}{6678}$	33	
June 27, 1956	Q	3	,,	57	
Mar. 24, 1967	Q	3	,,	53	
					36·5
Sept. 16, 1958	Q	4	$\frac{4471}{4922}$	15	
Mar. 23, 1959	Q	4	,,	30	
Jan. 16, 1965	Q	4	,,	31	
June 14, 1956	A	3	$\frac{5876}{6678}$	3·3	
Dec. 19, 1956	A	4	$\frac{4471}{4922}$	2·8	
Dec. 14, 1957	A	4	,,	6·6	
					4·0
Aug. 2, 1958	A	4	,,	2·0	
Mar. 23, 1959	A	4	,,	4·1	
Sept. 26, 1963	A	4	,,	5·1	

quiescent objects. However, when we go to active prominences, we find that the triplet/singlet line intensity ratios are greatly decreased, approaching the theoretical 3:1 value. For the triplet/singlet line intensity ratio to approach 3:1, the 2^1P level must not be significantly depopulated relative to the 2^3P level. There are basically three ways in which this may be achieved, viz.

(1) the resonance line, 584 Å, is optically thick,
(2) the 2^1P (and the 2^3P) level is populated mainly by collisions, or
(3) He shows a recombination spectrum.

Rough estimates indicate that the first alternative may be realized for densities of He I of the order of 10^8 cm^{-3}. We then have for the opacity in the resonance line $\tau_{584} = n(\text{He I}).L\alpha(584) \approx 10^8$ cm^{-3} 10^8 cm 10^{-13} cm$^2 = 10^3$. To accomplish this one would require that the total He density be about 10^{11} cm^{-3} and furthermore that a fraction 10^{-3} of the He be neutral. The latter condition holds for temperatures up to $T_e \approx 40000$ K (Athay and Johnson, 1960). According to Zirker (1959) He II becomes less abundant than He III for $T_e > 54000$ K.

In the cool plasma of quiescent prominences the total He density may be less than 10^{11} cm^{-3}, in which case we would not expect the 584 Å line to be as optically thick as required. However, this is a borderline situation, and more work is needed to settle the question. In active prominences it is more likely that we are approaching the necessary conditions ($\tau_{584} \approx 10^3 - 10^4$) to realize alternative 1.

Alternative 2 would be realized for a sufficiently high electron density in the prominence. The question is merely whether the necessary densities are reached in active prominences, in quiescent objects $n_e \approx 10^{11}$ cm^{-3} (Equation (II.23)) which is too low. Calculations (Athay and Johnson, 1960) indicate that radiative excitation dominate the collisional rates for $n_e < 10^{12}$ cm^{-3}, and even for higher values. Hence, it seems doubtful that sufficient densities are found in most active prominences for alternative 2 to be realized. In flares, however, where n_e may reach values of 10^{13} cm^{-3}, the situation may be different.

Alternative 3 applies at high temperatures and has been invoked by Zirin and Acton (1967). The possibility that lines from He I are formed by cascade following recombination points to some interesting questions regarding the physics of very active prominences. We shall pursue this further in Section b below.

(b) *Ionized Helium*

Zirin (1964) has shown that the He II spectrum in very active prominences may be a recombination spectrum. In the following we shall consider his arguments. Three lines of He II are observed in the visible part of the spectrum. They are the $n=4 \to n=3$ transition at 4686 Å, the $n=7 \to n=4$ transition at 5411 Å and the line 4541 Å due to the transition $n=9 \to n=4$, where n is the principal quantum number. If these lines are excited at a certain excitation temperature T_{ex}, we expect the ratio of the intensities of any two of the lines to be given by (see Equation (III.44))

$$\frac{I_1}{I_2} = \frac{g_1}{g_2} \frac{A_1}{A_2} \frac{e^{-\chi_1/kT_{ex}}}{e^{-\chi_2/kT_{ex}}}, \tag{III.45}$$

where g, A and χ are the statistical weight, the Einstein A coefficient and the excitation potential associated with the upper levels of the lines, respectively. Inserting appropriate values we find for the three lines under steady state conditions:

$$I(4686):I(5411):I(4541) = 75:4:1\,, \tag{III.46}$$

ignoring differences in χ from line to line. However, observations in active prominences give

$$I(4686):I(5411):I(4541) = 20:4:1\,. \tag{III.47}$$

On the other hand, if the He II spectrum is formed in cascade following recombination, the intensity of a line depends on the rate of electrons arriving in the upper level of the line (since this rate is much lower than the Einstein A coefficient for transition out of the level). Under these conditions Zirin found the following ratios:

$$I(4686):I(5411):I(4541) = 30:5:1\,,$$

which are in fair agreement with observations, Equation (III.47), and not with the values expected from steady state conditions, Equation (III.46).

3.2.3. Active prominences as magnetic flux tubes

When one looks at the shape of many active prominences, often it is difficult to avoid the impression that we observe the outlines of magnetic flux tubes. The most convincing examples of this are furnished by loop prominences (see Figure III.7) but arch

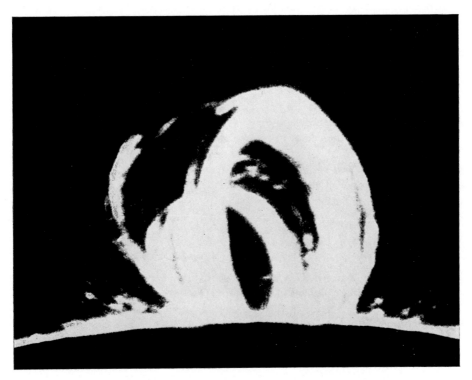

Fig. III.7. Loop prominence system (courtesy Sacramento Peak Observatory).

filaments, fibrils and many active region filaments probably also can be treated as flux tubes. To study the geometry of such tubes it is helpful to compare with the configuration of potential field lines. In the next section, therefore, we first discuss briefly how one calculates potential magnetic field configuration in the corona, given certain information on the field at photospheric or chromospheric levels.

(a) Potential Field Calculations

Schmidt (1964, 1965, 1966), Godovnikov and Smirnova (1965), and Semel (1967) have developed programs to study the coronal field configuration, given information on the field at the surface, under the assumption that no current flows in the coronal regions.
Under such conditions, since

$$\nabla \times \mathbf{B} = 0 \,,$$

the field can be derived from a scalar function ϕ

$$\mathbf{B} = \nabla \phi \qquad \qquad \text{(III.48)}$$

$\mathbf{B}$ is solenoidal, and the magnetic scalar potential ϕ satisfies Laplace's equation:

$$\nabla^2 \phi = 0 \,.$$

The currents responsible for the field B flow in the photosphere, and if we can specify ϕ on this boundary, we have the boundary value problem of finding ϕ (or $\mathbf{B}$) everywhere above the boundary. From observations we know either values of ϕ (leading to a Dirichlet problem) or of the normal component of the potential, $\partial \phi / \partial z$ (leading to a Neumann problem).

Let P indicate the photospheric surface, considered plane, in which the currents flow. Then the unique computation of ϕ above P is given by (Stratton, 1941)

$$\phi(x, y, z) = \frac{1}{2\pi} \int_P \frac{1}{r} \frac{\partial \phi}{\partial z} \, da \,, \qquad \qquad \text{(III.49)}$$

where da is an element of the plane P.

When we observe at the center of the disk, $B_z = B_\parallel$, i.e., the observed longitudinal component is equal to the vertical component, and we specify $\partial \phi / \partial z$ directly in Equation (III.49). When we observe away from the center of the solar disk, the observed $B_\parallel$ no longer gives us B_z. Semel's analysis remedies this situation and Equation (III.49) is replaced by

$$\phi(x, y, z) = \int_\infty^{x, y, z} B_\parallel \, dl \,, \qquad \qquad \text{(III.50)}$$

where integration is along the line of sight from the observer at infinity where the potential is zero. For practical use of Equation (III.50) see, for example, Harvey (1969) who computed potential fields in the corona near many prominences using data on observed values of $B_\parallel$ in the photosphere (the plane P).

Along field lines

$$\mathbf{B} \times d\mathbf{s} = 0 \,, \qquad \qquad \text{(III.51)}$$

where $d\mathbf{s}$ is an element along a field line. From Equation (III.51)

$$\frac{\mathrm{d}x}{B_x} = \frac{\mathrm{d}y}{B_y} = \frac{\mathrm{d}z}{B_z} = c_i \,, \tag{III.52}$$

where the constant c_i characterizes the particular line numbered i, see Equation (III.43).

The technique outlined above will give reasonable results only if the magnetic field is constant within each elementary area of P. Usually, the elementary area is given by the spatial resolution of the magnetograph providing the observed $B_{\parallel}$.

(b) *Loop Prominence Systems*

Well-developed loops are among the most spectacular prominence phenomena observed. They are rooted in a flare region or in a sunspot umbra (Bruzek, 1964b). The material seen streaming down the two legs of the loop presumably delineates the lines of force of a magnetic field joining one flare region and another, or one flare region and a sunspot. Harvey (1969) has shown that the orientation of loops coincides well with magnetic field lines computed with potential theory Equation (III.50), see Figure III.8. However, the observed field strength is much greater than predicted by a potential field. Part of this discrepancy may be explained in terms of the fine structure of the loops. Even though the overall shape of the loop system coincides well with field lines computed with potential theory, an individual loop considered as a flux tube, seems to have a cross section, πr^2, that does not increase rapidly with height. The extreme case of a constant cross-section would imply $B\pi r^2 \approx$ constant, and would indicate considerable field strength at great heights. The reason for this behavior of the loops is not clear, but may indicate considerable twist of the field lines.

Gold (1964, 1968, 1971) has advocated the view that since the prominence plasma is a good conductor, one should expect force-free fields, and not a potential field, in structures like loop prominences. Such fields would lead to helical configurations, and one would expect much higher field strengths at the tops of loops than with a potential field.

According to Gold the notion of helical fields is not in contradiction with observations, since it would be very difficult to observe the twists in bundles of field lines.

(c) *Arch Filament Systems*

Arch filament systems recently have been studied extensively by Bruzek (1967, 1968, 1969a), but already Waldmeier (1937) and Ellison (1944) noted their presence in young active regions, see also Bumba and Howard (1965) and Martres *et al.* (1966). Since the filaments connect areas of opposite polarity, they cross the neutral line in the longitudinal magnetic field, and appear to follow the field lines. The filaments are associated with the formation of sunspots and are often seen to rise into the corona. What we observe may therefore be emerging magnetic flux tubes, seen to rise and expand as the sunspot group grows (see also Kleczek, 1968; Zirin, 1972). The orientation of arch filaments is in excellent agreement with magnetic field lines computed from potential theory (Equation (III.50)) (Harvey, 1969).

(d) *Active Region Filaments*

While fibrils, as flux tubes, behave very similar to arch filaments, this is not the case for active region filaments. Prominences of the latter type do connect opposite

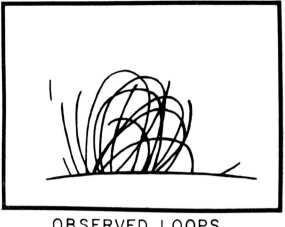

OBSERVED LOOPS

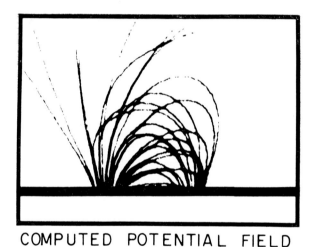

COMPUTED POTENTIAL FIELD

Fig. III.8. Comparison of loop prominence system with computed potential field lines (courtesy J. W. Harvey).

magnetic polarities, but most of their length coincides with – or lies above – the neutral line, suggesting considerable shear of the magnetic configuration. In all three types material motion is along the long axis of the prominences, from one end to the other in active region filaments (S. F. Smith, 1968), from the top down the two legs in arch filaments. Hence they may all be considered magnetic channels along which matter is allowed to flow, but there the similarities seem to end. For instance, the shape of active filaments compares poorly with field-line configurations computed from potential theory (Harvey, 1969). This is not really surprising. Such filaments are found in regions of often great activity, and significant electric currents probably flow in and above these regions (Severny, 1965). The use of potential field theory then becomes untenable, and more complicated conditions must be studied.

Raadu and Nakagawa (1971) have considered the next simplest case, that of force-free field configurations. From Equations (III.21) and (III.22) and the assumption that the field falls off exponentially with height, i.e.,

$$\mathbf{B} = \mathbf{B}_0 \, e^{-z} \,, \tag{III.53}$$

Raadu and Nakagawa found solutions representing force-free fields near bipolar regions. They showed that one may construct field lines that are similar to the shape of commonly observed active filaments.

CHAPTER IV

FORMATION OF PROMINENCES

In this chapter we shall consider the formation of prominences. The different processes may lead to stable or unstable objects, but we postpone the discussion of stability of prominences to Chapter V. In their prominence classification Menzel and Evans (1953) realized the important distinction between objects forming from above and objects forming from below (see Section 1.3.3). We refer to the former case as condensation, and for the latter we use the term injection.

A number of authors have advocated the view that many types of prominences form from material condensing out of the corona (Kiepenheuer, 1953b, 1959; Parker, 1953; Zanstra, 1955; Kleczek, 1957, 1958; Lüst and Zirin, 1960; Uchida, 1963; Doherty and Menzel, 1965; Field, 1965; Hunter, 1966; Raju, 1966, 1968; Kuperus and Tandberg-Hanssen, 1967; Olson and Lykoudis, 1967; Nakagawa, 1970; Goldsmith, 1971; Hildner, 1971).

On the other hand, there obviously are prominences which originate from below, being shot out of photospheric or chromospheric layers. Foremost in this group are surges and sprays. Also spicules seem to have a similar origin. No comprehensive theory exists to explain in detail the formation of these injections, but several interesting aspects of the processes that might be involved have been discussed (Schlüter, 1957b; Jensen, 1959; Gopasyuk, 1960; Warwick, 1962; Jefferies and Orrall, 1965; Pikel'ner, 1971).

4.1. Condensations

4.1.1. CONDENSATION AS A THERMAL INSTABILITY

If the thermal equilibrium of a plasma at temperature T_c and density n_c (to be identified with the corona) is unstable, formation of a condensation of lower temperature, $T_p < T_c$, and higher density, $n_p > n_c$ (to be identified with a prominence), becomes possible. Field (1965), in a thorough analysis of this problem, points out that Weyman (1960) seems to be the first to give the correct instability criterion for the formation of such a condensation, even though Parker (1953) did much of the pioneering work. Zanstra (1955), Kleczek (1958), and others have also contributed to the early discussion. Field analyzed the different possible instability criteria in an infinite uniform medium in the absence of magnetic fields, using linearized equations. In such a medium the equation of continuity

$$\frac{\partial \rho}{\partial t} + \nabla \cdot \rho \mathbf{v} = 0 \,, \tag{IV.1}$$

and the equation of motion

$$\rho \frac{d\mathbf{v}}{dt} + \nabla p = 0 \tag{IV.2}$$

are automatically satisfied. It is in the choice of the form of the energy equation

$$\frac{\partial \varepsilon}{\partial t} + \nabla \cdot \varepsilon v = 0 , \tag{IV.3}$$

where ε is the energy density of the plasma, that most treatments differ. Field used a heat equation of the form

$$\frac{dp}{dt} - \gamma \frac{p}{\rho} \frac{d\rho}{dt} - \rho(\gamma - 1)(G - L) = 0 , \tag{IV.4}$$

where G and L are the energy gains and losses respectively per unit mass and per second, exclusive of thermal conduction.

The infinite plasma, e.g., the corona, may be perturbed (in density and temperature), but in such a way that the pressure remains constant. If we assume the validity of a perfect gas law, such an isobaric perturbation will lead to instability if

$$\left[\frac{\delta(G - L)}{\delta T} \right]_p > 0$$

or

$$\left[\frac{\delta(G - L)}{\delta T} \right]_\rho - \frac{\rho_c}{T_c} \left[\frac{\delta(G - L)}{\delta \rho} \right]_T > 0 . \tag{IV.5}$$

As Field points out, this criterion is consistent with the equation of motion, and condensations are governed by this isobaric criterion. A generalized criterion, valid in the presence of thermal conduction, is given by Hunter (1970).

The less restricted isochoric perturbations lead to an instability criterion

$$\left[\frac{\delta(G - L)}{\delta T} \right]_\rho > 0 \tag{IV.6}$$

and is the one given by Parker (1953). However, this criterion is not consistent with Equation (IV.2), since pressure variations due to the allowed temperature variations will generate motions that can lead to non-constant density.

If entropy, s, instead of pressure, is kept constant during the perturbations, we are faced with an isentropic perturbation, and the corresponding instability criterion is

$$\left[\frac{\delta(G - L)}{\delta T} \right]_s > 0$$

or

$$\left[\frac{\delta(G - L)}{\delta T} \right]_\rho + \frac{1}{\gamma - 1} \frac{\rho_c}{T_c} \left[\frac{\delta(G - L)}{\delta \rho} \right]_T > 0 . \tag{IV.7}$$

The isentropic criterion applies to conditions governed by adiabatic motions, i.e., to sound waves, and has been studied by Hunter (1966), while non-adiabatic motions are involved in the condensation mode governed by criterion (IV.5).

Most of the linear treatments of the condensation process have considered only the energy equation (Equation (IV.3)) in one form or another. For a thorough derivation of the equations for the conservation of energy see Ledoux and Walraven (1958). However, the heat equation appropriate to the condensation problem is nothing more than an expression of the first law of thermodynamics

$$\frac{dU}{dt} = \frac{dQ}{dt} + \frac{p}{\rho^2}\frac{d\rho}{dt} , \tag{IV.8}$$

where U is the internal energy and Q the energy fed into the plasma from external sources. With a perfect gas law, Equation (IV.8) may be written

$$\frac{dp}{dt} - \gamma\frac{p}{\rho}\frac{d\rho}{dt} - (\gamma-1)\rho\frac{dQ}{dt} = 0 . \tag{IV.9}$$

The rate at which energy is fed into the plasma, dQ/dt, is the net gain, i.e., the difference between all gain terms, G, and all losses, L, compare Equation (IV.4).

Different authors have included different terms in G and in L, G_{rad} = absorbed radiation energy, G_{comp} = heating due to compression by external forces, G_{mech} = dissipation of mechanical energy, L_{rad} = loss by emission of radiation, and the effect of thermal conduction. Thermal conduction may lead to a loss or a gain depending on the temperature gradient in the condensing preprominence. For example, if compression temporarily has raised the temperature above the ambient coronal temperature, conduction losses occur (Lüst and Zirin, 1960; Olson and Lykoudis, 1967). Later in the cooling process, the temperature gradient will have changed sign, and conduction will try to destroy the condensation. It is convenient to single out thermal conduction and write Equation (IV.9) in the form

$$\frac{dp}{dt} - \gamma\frac{p}{\rho}\frac{d\rho}{dt} - \rho(\gamma-1)(G-L) - \rho(\gamma-1)\nabla\cdot(K\nabla T) = 0 , \tag{IV.10}$$

where K is the coefficient of thermal conductivity.

If the gain terms include external compression effects, and the losses are due to radiation, i.e.,

$$G-L = G_{comp} - L_{rad} , \tag{IV.11a}$$

we retrieve the form used by Kleczek (1957). Lüst and Zirin used

$$G-L = G_{comp} - L_{rad} - \nabla\cdot(K\nabla T) , \tag{IV.11b}$$

i.e., as stated above, they retained the conduction term $\nabla\cdot(K\nabla T)$ as a loss term, since in their case the plasma became heated relative to the ambient corona due to compression.

Kuperus and Tandberg-Hanssen (1967) included the terms

$$G-L = G_{mech} - L_{rad} \tag{IV.12}$$

and neglected the effect of thermal conduction because of the presence of magnetic fields. This form (Equation (IV.12)) is also the one used in the more sophisticated treatments due to Raju (1968), Nakagawa (1970), and Hildner (1971).

The exact form of the energy gain function, G_{mech}, is not obvious. If the corona is heated mainly by the dissipation of wave energy from below, it is likely that G_{mech}

will depend linearly on the local density (Weyman, 1960; Uchida, 1963), and we may write

$$G_{\text{mech}} = C_1 \rho , \tag{IV.13}$$

where C_1 is a constant to be specified at thermal equilibrium. Equation (IV.13) should hold also for heating due to corpuscular streams (Raju, 1966). However, more work needs to be done on the functional form of G_{mech}. The radiation losses, i.e., the divergence of the radiative flux $\mathbf{F}_{\text{rad}}$, $L_{\text{rad}} = -\nabla \cdot \mathbf{F}_{\text{rad}}$, are due to emission in lines (bound-bound transitions) as well as in continua (free-free and free-bound transitions), i.e.,

$$L_{\text{rad}} = L_{\text{ff}} + L_{\text{bf}} + L_{\text{bb}} .$$

The loss terms have been calculated by Orrall and Zirker (1961), Hirayama (1964), Doherty and Menzel (1965), Pottasch (1965), Raju (1968), and Cox and Tucker (1969).

Building on a large number of investigations, we seem to be able to approximate L_{rad} adequately by the form

$$L_{\text{rad}} = C_2(T)\rho^2 T^{\alpha(T)} , \tag{IV.14}$$

where the temperature-dependent coefficients $C_2(T)$ and exponents $\alpha(T)$ can be considered constants within certain temperature ranges (Nakagawa, 1970; Goldsmith, 1971; Raadu, 1971; Hildner, 1972) (see Figure IV.1). We notice that the loss function is small for temperatures below a few thousand degrees, reaches a broad maximum between approximately 5×10^4 and 3×10^5 K and then decreases again.

Extensive quasi-linear (Nakagawa, 1970) and non-linear treatments (Raju, 1966, 1968; and Hildner, 1971) of the condensation process have been presented, using the hydromagnetic equations for a perfect electric conductor. A macroscopic treatment is applicable since the crucial characteristic length in a plasma pervaded by a magnetic field is the gyration radius. Even a weak field of 10^{-3} G gives gyration radii of less than 10 km in the corona.

When one includes the effects of a magnetic field $\mathbf{B}$, it is possible to shield an initial temperature perturbation from the otherwise overwhelming heat conduction that very quickly would destroy any embryonic condensation (Rosseland et al., 1958). This inclusion leaves the equations of mass and heat conservation unchanged (Equations (IV.1) and (IV.10)), while the equation for the conservation of momentum becomes

$$\rho \frac{d\mathbf{v}}{dt} + \nabla p - \rho g - \frac{1}{c}\mathbf{j} \times \mathbf{B} = 0 . \tag{IV.15}$$

If a perfect gas law is assumed

$$p = \rho R T ,$$

the set of equations becomes closed with the addition of Maxwell's equations and the assumption of infinite electric conductivity, i.e.,

$$\nabla \times \mathbf{B} - \frac{4\pi}{c}\mathbf{j} = 0 \tag{IV.16}$$

$$\nabla \cdot \mathbf{B} = 0 \tag{IV.17}$$

$$\frac{1}{c}\frac{\partial \mathbf{B}}{\partial t} + \nabla \times \mathbf{E} = 0 \tag{IV.18}$$

and

$$E = -\frac{1}{c}\mathbf{v} \times \mathbf{B},$$
(IV.19)

or

$$\frac{\partial \mathbf{B}}{\partial t} - \nabla \times (\mathbf{v} \times \mathbf{B}) = 0.$$
(IV.19')

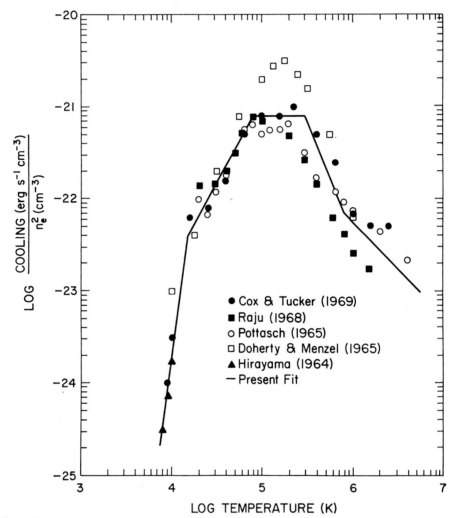

Fig. IV.1. Radiation losses as function of temperature from tenuous solar plasma (Equation (IV.14)). The parameters $C_2(T)$ and $\alpha(T)$ in Equation (IV.14) that describe the solid line fit to the points are given in the following table, due to Hildner (1972):

$T\,(\mathrm{K})$	$C_2(T)$	$\alpha(T)$
$8 \times 10^5 < T$	$5 \cdot 5 \times 10^{-17}$	-1
$3 \times 10^5 < T < 8 \times 10^5$	$3 \cdot 9 \times 10^{-8}$	$-2 \cdot 5$
$8 \times 10^4 < T < 3 \times 10^5$	$8 \cdot 0 \times 10^{-22}$	0
$1 \cdot 5 \times 10^4 < T < 8 \times 10^4$	$1 \cdot 2 \times 10^{-30}$	$+1 \cdot 8$
$T < 1 \cdot 5 \times 10^4$	$4 \cdot 7 \times 10^{-54}$	$+7 \cdot 4$

4.1.2. Condensation of Prominences

The interplay of the gain and loss functions indicates that the plasma is thermally unstable against isobaric condensation for certain behavior of the loss-function. However, this does not mean that the whole corona should collapse into prominences. Even in an otherwise unstable temperature regime, a local temperature perturbation will quickly be evened out due to thermal conductivity. As mentioned above, only where the magnetic field can shield such an embryonic condensation, is the process likely to continue. Then the negative value of $G - L$ leads to cooling, the pressure drops, material will be drawn into the beginning condensation where it will be cooled, and the condensation process proceeds.

However, there may be local minima in the loss curve of Figure IV.1 at coronal temperatures and below. Under such conditions one may expect stabilizing effects to take over and prevent condensation. This possibility does not seem to have been explored in the literature.

Also, even for an unstable medium, the time scale involved will often be too long. This means that, at best, unforced condensation may produce quiescent prominences, but the rapid condensation necessary to explain some active objects requires an additional forcing mechanism. It turns out that if one could first compress the plasma, the resulting cooling would then go much quicker, and this may be a possible way out.

(a) Quiescent Prominences

The importance of the magnetic field configuration for the formation of quiescent prominences was considered in the model by Kuperus and Tandberg-Hanssen (1967). Most quiescent prominences are associated with characteristic coronal streamers that appear above them, and on either side of the filament one finds, in the chromosphere, magnetic plages of opposite polarity. A model for the history of the corresponding coronal field structure is shown schematically in Figure IV.2. In the pre-active phase of the active region Figure IV. 2a shows a closed dipole-type configuration. The corona overlying the center of activity is more intensely heated than the quiescent corona

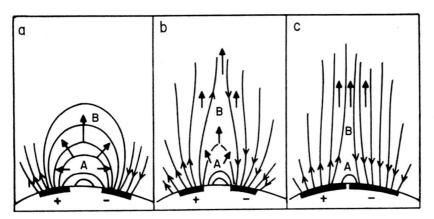

Fig. IV.2. The possible magnetic field configuration over an active region during (a) the early phase of activity, (b) the main phase of activity, (c) the post-active phase.

(Kuperus, 1965), whence the gas pressure in region A becomes higher than at the same level outside the active region. The active coronal region may be heated until, at some height B, matter is blown out, thus opening the field lines (Figure IV.2b), the safety-valve mechanism (Parker, 1963a). In region B the field structure is determined by the outward flow, while region A possesses a more or less potential field. Between A and B we find a region where the magnetic energy is comparable to the kinetic energy of the solar wind, and the field configuration there is not easily described.

During the post-active phase (Figure IV.2c) the enhanced coronal expansion ceases, and the field lines are no longer pushed sideways. The region near the plane of symmetry has very weak fields, whence matter is compressed there to maintain lateral equilibrium. The material near this neutral sheet has higher density than the surrounding corona and loses more energy by radiation. The magnetic field is radial and G_{mech} is assumed to be the same as in the surrounding corona. Hence it is not possible to compensate for the enhanced radiation loss, and the neutral-sheet region becomes thermally unstable.

Kuperus and Tandberg-Hanssen assumed that the magnetic field configuration and the gas pressure in the corona near the neutral sheet do not change during the condensation, and that thermal conduction is sufficiently inhibited to be neglected. Equations (IV.9) and (IV.12) may then be written

$$c_p \frac{\partial T}{\partial t} = G_{mech} - L_{rad} ,$$
(IV.20)

where $c_p = \gamma R/(\gamma - 1)$ is the specific heat at constant pressure. For the radiation losses they used Orrall and Zirker's (1961) expression, which gives too small a value, $L_{rad} = 6 \cdot 3 \times 10^{24} [\rho(z, t)]^2$ erg cm^{-3} s^{-1}. The heat supply by dissipation of mechanical energy just balances the radiative losses of the region far away from the sheet, i.e., $G_{mech} = 6 \cdot 3 \times 10^{24} [\rho_c(z)]^2$. The solution of Equation (IV.20) indicates that the temperature of the neutral sheet (the pre-prominence) drops by a factor of 10 and the density increases by the same factor, in about 10^5 s from an initial coronal temperature of $T_c = 10^6$ K.

After the condensation has proceeded for some time, tearing-mode instabilities may set in to connect field-lines across the neutral sheet. This would cause a filamentary fine structure in the forming prominence, which is generally observed. Also, the lower-lying field-lines would become more or less horizontal and provide support for the denser prominence material (see Figure IV.3).

The model is crude, with the thermal instability inadequately treated, but it links the formation of quiescent prominences to basic phenomena in the corona, related to several well-observed characteristics.

The strong coupling of the system of Equations (IV.1), (IV.10), (IV.15) to (IV.19) makes the general, non-linear treatment mathematically intractable. Hence, one omits thermal conduction and assumes a uniform magnetic field of unrealistically simple configuration.

Raju (1966, 1968) considered a magnetic field parallel to the long axis of a cylindrical configuration in the corona. He let the cylinder undergo a radial compression and showed that instability will set in and lead to a condensation for sufficiently weak initial fields. It is not obvious how a radial compression can arise, and the difficulty associated with this initial compression, which, in some form or another also is crucial

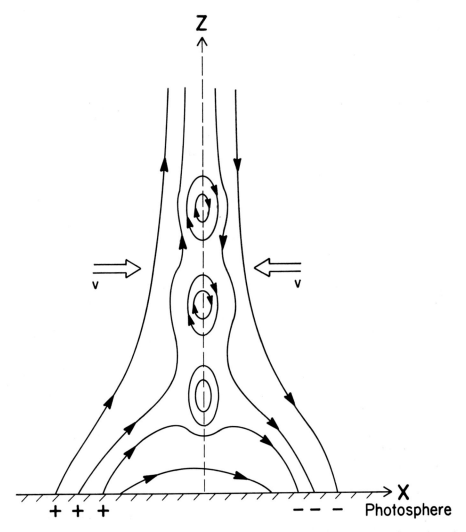

Fig. IV.3. Schematic representation of the magnetic field configuration during the formation of a quiescent prominence in Kuperus and Tandberg-Hanssen's (1967) model.

to other treatments of the condensation problem, is not solved. Raju showed that prominence conditions may be reached after about 10^6 s, i.e., 10 to 12 days, which seems too long. Also, Raju could not produce prominences with magnetic field strengths in excess of 0·1 G, which is too low. Nevertheless, his work constitutes an important step forward, considering, as it does, the effects of the intensification of the magnetic field as the field lines are swept into the condensation. The results are encouraging considering the simplified model used.

The most complete treatment of the non-linear condensation process is due to Hildner (1971, 1972). He considered an initial horizontal magnetic field, and for computational reasons recast the energy equation in the form

$$\frac{\partial \varepsilon}{\partial t} + \nabla \cdot \mathbf{S} - (G - L) = 0 , \tag{IV.21}$$

where the total energy density, ε, is

$$\varepsilon = \frac{p}{\gamma - 1} + \frac{1}{2}\rho v^2 + \frac{B^2}{8\pi} + \rho g z$$

and the energy flow, $\mathbf{S}$, the Poynting vector, is

$$\mathbf{S} = \mathbf{E} \times \mathbf{B} + \left(\frac{\gamma p}{\gamma - 1} + \frac{1}{2}\rho v^2 \right) \mathbf{v} .$$

Hildner numerically solved the system of Equations (IV.1), (IV.15) to (IV.19), (IV.21), both for one- and two-dimensional cases, and showed that an initial density perturbation will grow to a condensation and will be supported against gravity by the magnetic field. Instabilities in the computation prevented him from reaching genuine prominence conditions, but the trend was clearly established.

Kuperus and Tandberg-Hanssen (1967) and Raju (1968) assumed all variables inside the condensation to be spatially uniform (see also Uchida (1963)), and this uniformity permitted the fluid motions to be neglected. Hildner included the fluid dynamics of the problem, and showed that the response of the fluid to a local density increase abruptly changes character when the initial sound speed exceeds the Alfvén speed. If the flow is small and perpendicular to the magnetic field $\mathbf{B}$, the system of equations is hyperbolic. If the flow is at an arbitrary angle to $\mathbf{B}$, the character of the governing equations depends strongly on the value of this angle, and has mixed features of both hyperbolic and elliptic equations. This leads to formidable mathematical problems. In many of these instances it is not known how to formulate the function giving the shape of the characteristics in solution space (Chu and Grad, 1965). However, even though much work remains, Hildner has pushed the non-linear treatment of the condensation process an important step further.

(b) Active Prominences

A striking feature of many active regions is the continual appearance of loops and of 'coronal rain'-type prominences. They seem to condense out of the overlying corona in order of minutes.

Kleczek (1958) realized that a simple thermal instability would not produce the rapid condensation necessary to explain the formation of such active prominences. However, since the energy loss by radiation is proportional to density squared, he argues that if the coronal plasma first could be compressed, the subsequent cooling time then might be short enough. In Kleczek's model, it is hot coronal regions that are being formed and that thereafter cool, according to Equation (IV.11a).

If the compression is very rapid, the radiation loss will be small, very little energy will be lost by thermal conduction, and the process will be quasi-adiabatic,

$$n^{1-\gamma} \cdot \kappa T = \text{const.} \quad \text{or} \quad T \propto n^{2/3} \quad \left(\gamma = \frac{5}{3} \right) .$$

This may be written in differential form,

$$dT = \frac{2}{3}\frac{T}{n}\, dn\,. \tag{IV.22}$$

Kleczek determined the constant by inserting values of n and T for the initial condition of the plasma, $n_c = 10^8$ cm^{-3} and $T_e = 10^6$ K. Then, $T = \sqrt[3]{100 n^2}$. Since the pressure changes approximately as $n^{5/3}$, we find $p \approx T^{5/2}$. In the limiting case of adiabatic compression, the time over which compression takes place must be infinitesimal to prevent escape of energy from the volume.

On the other hand, if the compression is quasi-isothermal, $T = T_e = 10^6$ K. The change in energy is $dU = 3/2 nk\, dT + 3/2 kT\, dn$. In isothermal compression, the energy $3/2 nk\, dT$ is radiated away, and this is given by Equation (IV.14). However, Klezcek used a simpler expression for the radiation losses, namely

$$L_{rad} = n^2 \left[aT^{1/2} + \frac{b+c}{T^{1/2}} \right],$$

where the coefficients a, b and c give the contribution due to free-free, free-bound and bound-bound radiation, respectively (Kleczek, 1957). To find an expression for dT, he considered the isothermal compression as a series of infinitely small adiabatic changes. The energy corresponding to the increment dT is then emitted as radiation according to

$$\frac{3}{2}nk\, dT = n^2 \left[aT^{1/2} + \frac{b+c}{T^{1/2}} \right] dt\,. \tag{IV.23}$$

Equations (IV.22) and (IV.23) give

$$dt = \frac{kT}{n^2[aT^{1/2} + (b+c)T^{-1/2}]}\, dn\,,$$

which, when integrated, gives the time during which the hypothetical compression operates to compress the plasma to the limiting density. Inserting the values $a = 2 \times 10^{-27}$, $b = 3 \times 10^{-21}$, and $c = 2 \times 10^{-20}$ corresponding to a million degree corona, we find with Kleczek for a pressure increase of a factor of 100:

$$t \approx 2 \times 10^{13} \int_{10^8}^{10^{10}} dn/n^2 \approx 2 \times 10^5 \text{ s}\,.$$

This estimate shows that isothermal compression is too slow, and Kleczek's model calls for compression intermediate between the extremely fast adiabatic and the too-slow isothermal compressions.

Kleczek's model requires a strong initial compression ($n \approx 10^{11}$ cm^{-3}) to produce the subsequent rapid cooling. However, Lüst and Zirin (1960) pointed out that the prominences in question seem to condense out of a plasma that already is hotter than normal ($T_c \geq 2 \times 10^6$ K). If one compressed such a plasma, the temperature would probably become much too high before the cooling started. It was to remedy this defect that they considered how thermal conduction may initially cool the condensation. We shall follow their arguments, but remedy some mistakes (see Shklovsky, 1965). They assumed that an element of coronal plasma of temperature T_c and density

n_c, contained in a magnetic flux tube, is subjected to a strong uniform compression perpendicular to the magnetic field according to the equation

$$n = n_c(1 + \alpha t) . \tag{IV.24}$$

Their heat equation is Equation (IV.11b). If the collision times for the particles are short, the ratio of specific heats may be taken as $\gamma = 5/3$, $T/T_c = (1 + \alpha t)^\gamma$ and

$$G_{comp} = \frac{3}{2} n k T_c \alpha \gamma (1 + \alpha t)^{\gamma-1} . \tag{IV.25}$$

The magnetic field will prevent heat conduction across the lines of force; thus heat is only conducted out of the element of plasma in the direction of the field. The expression for this loss is (Spitzer, 1956)

$$L_{cond} = \text{const} \frac{\varepsilon \delta_T}{Z \ln \varLambda} T^{5/2} \nabla T , \tag{IV.26}$$

where ε, δ_T and $\ln \varLambda$ are given by Spitzer.

Lüst and Zirin approximated Equation (IV.26) by

$$L_{cond} = \text{const } T^{5/2} \frac{\varDelta T}{L^2} , \tag{IV.26'}$$

where L is the length of the flux tube and $\varDelta T$ the overall temperature difference. Further, they approximated the radiation loss by an expression similar to the one used by Kleczek. With the expression for L_{rad} and inserting Equations (IV.24), (IV.25) and (IV.26) in Equation (IV.11b), we arrive at an equation that may be integrated for a number of starting values of n_c and T_c, using α as a parameter. Lüst and Zirin found that the temperature will first increase to a maximum, T_{max}, at a time, t_{max}, due to the initial compression, after which the increased internal energy is radiated and conducted away and the cooling proceeds. The cooling time, t_{cool}, may be defined as the time it takes for the plasma to cool to a temperature $T < 2 \times 10^5$ K. The higher the compression rate, α, the faster the initial temperature increases, but the subsequent cooling is also faster. By proper choice of α and for T_c in the range $1-2 \times 10^6$ K and $n_c = 10^9 - 10^{10}$ cm^{-3}, t_{max} is of the order of $\frac{1}{2}$ h or less, while t_{cool} may be an hour or less, which seems to be a reasonable value for the time necessary to form active prominences. However, the model does not provide any explanation for the forces responsible for the compression. It is assumed that they may be of magnetic origin, but it is not obvious how this should work. The same criticism applies to Kleczek's model.

Some mistakes in Lüst and Zirin's paper were pointed out by Shklovsky (1965), and remedied by Olson and Lykoudis (1967). The work of the latter authors may be considered an extension of the Lüst and Zirin model, pertaining primarily to loop prominences.

Olson and Lykoudis found that they could account only for the condensation of lower loops in a loop prominence system. In this model the mass is supplied from a coronal condensation by compressing the flux tubes that presumably lead from the condensation and form the loops lower down. For loops higher than about 30000 km Olson and Lykoudis found that there will not be sufficient mass in the corona to account for a condensation process, and formation by injection is stipulated (see Section 4.2).

A somewhat different approach was taken by Goldsmith (1971) who analyzed the thermal effects in a post-flare plasma out of which the loop system was assumed to condense. He started out with a dense ($n_e \approx 10^{11}$ cm^{-3}) and hot ($T_e \approx 5 \times 10^6$ K) gas permeated by a magnetic field, and suggested that rapid cooling, due to radiative losses, leads to the formation of small denser and cooler knots in a time-scale of a few minutes. The development of a loop pattern is considered to be the result of the slower cooling of material in arched magnetic flux tubes at lower densities higher in the atmosphere. This behavior will lead to the observed effect that loops appear at greater and greater heights as the loop formation process proceeds. In this picture the upward expansion of the loop system (not the individual loop) is due to the propagation of a more or less spherical 'cooling wave', originating low in the dense flare plasma. The idea is interesting, but the treatment is too sketchy to allow definite statements as to its applicability to loops.

4.2. Injections

Models that describe prominences as being formed from below rely on an injection mechanism for the matter. This may be a macroscopic effect, carrying large blobs of plasma up from photospheric layers, or it may be a microscopic particle acceleration. Streams of charged particles seem at times, maybe always, to traverse coronal space, and the acceleration of such particles provides us with one of the most challenging problems of contemporary solar physics.

4.2.1. THE IMPORTANCE OF PARTICLE STREAMS

One of the best evidences for solar particle streams is the type III radio bursts. These events are ascribed to the passage from below of electrons (Kundu, 1965) or protons (D. Smith, 1970). The former alternative requires about 10^{35} electrons per burst in the $10^4 - 2 \times 10^5$ eV range, the latter needs about 10^{25} protons in the 5×10^7 eV energy range.

The ability of the Sun to eject energetic particles was conjectured from the solar cosmic ray events in February 1942 (Lange and Forbush, 1942). Acceleration of particles with cosmic ray energies ($> 10^7$ eV say) seems to be confined to flare regions, but less energetic particles come also from other active regions. For a survey of solar particle emissions see Fichtel and McDonald (1967).

Warwick (1962) has drawn attention to the importance of the 'particle aspect' of transient phenomena like flares. Instead of producing particle emission, the flares may be the result of the generation of fast particles in some other part of the Sun. Warwick assumes the particles to be accelerated in subphotospheric magnetic fields to sub-relativistic or relativistic velocities. No theory has been advanced to explain how particles can be accelerated in magnetic fields in these turbulent subphotospheric layers, but Warwick's contribution has been to emphasize the 'field and particle' picture as contrasted with more conventional approaches to flare models.

However, not only the generation of the fast particles but also their propagation presents serious difficulties if the particles are to be injected into prominences. The question is whether the high-energy particles have mean free paths of sufficient length to permit their transport. Following Warwick, we may state that the source of the particles can be no farther from the region of deposition than the distance in which protons of 3×10^8 eV would be stopped by ionization processes. The relation between the vertical range, h, of a 3×10^8 eV proton and its velocity, v, in a neutral H gas is (see Mott and Massey, 1949):

$$dh = -\frac{m_p m_e v^3}{4\pi e^4} \frac{dv}{n \ln (2m_e v^2/\chi(H))}, \tag{IV.27}$$

where $\chi(H)$ is the ionization energy of H and n the number density, given by $n = n_0 e^{-h/H}$, where H is the scale height of photospheric gases, that is, ≈ 110 km, and $n_0 \approx 10^{16}$ cm^{-3}. To integrate Equation (IV.27) from the initial velocity $v_0 \approx 0.8c$ to zero at height h_1, we assume that above h_1 there is a distance less than H where the atmosphere has the same density as at h_1. For $h > h_1$, Warwick assumes a negligible density. Integration of (IV.27) then gives

$$H e^{-h_1/H} = \frac{m_p m_e v_0^4}{16\pi e^4 n_0 \ln(2m_e v_0^2/\chi(H))} \approx 1.8 \times 10^9 \text{ cm}$$

or $h_1 \approx 480$ km, which is the greatest depth from which protons of energy 3×10^8 eV can emerge from the photosphere.

Warwick's picture offers interesting suggestions concerning certain low-lying flares, but for most prominences we seem to need an acceleration mechanism in the chromosphere or even in the corona.

4.2.2. SINGLE-PARTICLE ACCELERATION MECHANISMS

(a) *Introduction*

We know of only a few fundamental single-particle accelerating mechanisms for charged particles in nature, and they include the Swann mechanism, the Fermi mechanism, and the Sweet mechanism (Fichtel and McDonald, 1967). The first two mechanisms, which in many ways are the same process, are due to a varying magnetic field (we exclude here the possibility of acceleration in intrinsic electric potentials, since the solar atmosphere is highly electrically conductive). The Sweet mechanism relies on the annihilation of magnetic flux, and consequently consists of two parts if we want to apply it for acceleration purposes: first the theory of how magnetic fields may be destroyed and how the corresponding energy becomes available, and second how this available energy can be used to accelerate particles. We return to the Sweet mechanism in Section (c) below.

In the early 1930's Swann (1933) considered the effects of a changing magnetic sunspot field and showed that particle acceleration might occur. The mechanism of acceleration is similar to that applied in betatrons, and it is referred to both as the Swann mechanism and as betatron acceleration. As used by Swann the principle involved is an application of Faraday's law, which states that in the presence of a time-dependent magnetic field, an electromotive force is induced in a conductor. Let dl be an element of arc length of a closed loop bounding an open surface A. If a varying field **B** threads A, the electric field **E** induced in the loop satisfies the equation

$$\int_l \mathbf{E} \cdot d\mathbf{l} = -\frac{1}{c}\frac{d}{dt}\int_A \mathbf{B} \cdot d\mathbf{A},$$

where we have integrated Equation (IV.18) using Stokes' theorem.

Electrically charged particles, of charge Ze, will be accelerated in the electric field as they experience the force ZeE. This is the force that in betatrons accelerates electrons ($Z=1$) to velocities close to the velocity of light.

About 20 yr ago Fermi (1949, 1954) published the far-reaching theory for particle acceleration which bears his name. Basically the mechanism involves the relative motion of a magnetic region and the particle in question. Of special interest is the case where acceleration may occur when the charged particle is reflected repeatedly between two regions with strong magnetic fields (magnetic mirrors) moving toward each other. For such a periodic motion the integral of the particle momentum along a magnetic line of force through an entire period between the mirrors

$$J_\| = \frac{1}{2\pi}\oint p_\| \, ds \qquad\qquad (IV.28)$$

is constant, so long as the field does not change appreciably through one period, i.e., $J_\|$ is an adiabatic invariant, called the longitudinal invariant. Under these conditions the energy gain of the particle of mass m, between two mirrors whose distance originally was L, is

$$W = \frac{1}{2m}\left[p_\perp^2 + \left(\frac{L}{L-\Delta L}\right)^2 p_\|^2 \right],$$

when the distance has decreased by ΔL. The Fermi mechanism acts as the vehicle whereby part of the energy of a large number of particles – responsible for the motion of the magnetic fields – is transferred to a single particle. In Section b we shall consider the interrelation between the Swann and the Fermi mechanisms.

(b) *The Swann and Fermi Mechanisms*

In the introductory discussion of Swann's mechanism there are difficulties invoking conductors in the Sun's atmosphere, and besides, the coupling of the Swann and Fermi mechanisms is not apparent. To bring out the similarities we shall use a different approach and follow Hayakawa *et al.* (1964) in some detail.

Consider the motion of charged particles in a magnetic field in the guiding center approximation. The motion can be divided into three components, the gyration around the magnetic lines of force, characterized by the frequency $\omega_\perp$, the longitudinal motion along the lines, characterized by the frequency $\omega_\|$, and the drift motion over the surface of the magnetic flux tube, given by the frequency ω_D.

For periodic motions we introduce the angle variables ϕ, and the adiabatic invariants, or the action variables J. For the gyration

$$J_\perp = \frac{cp_\perp^2}{2ZeB} = \mu\frac{W}{Zec},$$

where $p_\perp = p \sin\alpha$, α is the pitch angle, μ the gyromagnetic moment, Ze the charge of the particle, and the total energy is $W = c\sqrt{m^2c^2 + p^2}$. We shall denote by v_B the velocity of the magnetic mirrors, and by $\mathbf{v}$ the velocity of the particle, relative to the observer. The angle variable, $\phi_\perp$, is the phase of the gyration. The second component, the longitudinal motion, is also often periodic, as for example when the particle is reflected repeatedly between two magnetic mirrors. Then $J_\parallel = 1/2\pi \oint p_\parallel \, ds$, as given in Equation (IV.28). Finally, for the drift motion component, J_D is the flux invariant, which reduces to the total angular momentum in an axially symmetric case. The frequencies of these three components are

$$\text{gyrofrequency } \omega_\perp = \frac{ZecB}{W}$$

$$\text{transit frequency } \omega_\parallel = 2\pi / \oint v_\parallel^{-1} \, ds$$

and

drift frequency ω_D, and they generally satisfy the inequalities

$$\omega_\perp > \omega_\parallel > \omega_D .$$

The energy of a particle is constant in a static magnetic field; hence time variations of the field $\mathbf{B}(s, t)$ are essential for acceleration. We have

$$\frac{dB}{dt} = \frac{\partial B}{\partial t} + v_B \frac{\partial B}{\partial s},$$

where v_B is the velocity of the magnetic region, and the energy of the particle in the observer's frame of reference is

$$\frac{dW}{dt} = \left(\frac{\partial W}{\partial t}\right)_s + \left(\frac{\partial W}{\partial t}\right)_F + \left(\frac{\partial W}{\partial t}\right)_I + \left(\frac{\partial W}{\partial t}\right)_T,$$

where the terms are,

$$\text{the Swann mechanism: } \left(\frac{\partial W}{\partial t}\right)_s = \frac{ZecJ_\perp}{W} \frac{\partial B}{\partial t} = \frac{c^2 p_\perp^2}{2WB} \frac{\partial B}{\partial t},$$

$$\text{the Fermi mechanism: } \left(\frac{\partial W}{\partial t}\right)_F = \frac{ZecJ_\perp}{W} v_B \frac{\partial B}{\partial s} = \frac{c^2 p_\perp^2}{2WB} v_B \frac{\partial B}{\partial s},$$

$$\text{the induction effect: } \left(\frac{\partial W}{\partial t}\right)_I = Zec \frac{\partial B}{\partial t} \sqrt{J_\perp(J_\perp + J_D)} \sin \phi_\perp$$

$$= ZecB \frac{\partial J_\perp}{\partial t},$$

$$\text{the transit-time effect: } \left(\frac{\partial W}{\partial t}\right)_T = -\frac{c^2 p_\parallel}{W} \frac{\partial}{\partial s} \left(\frac{1}{B} \frac{\partial B}{\partial t}\right) \sqrt{J_\perp(J_\perp + J_D)} \cos \phi_\perp,$$

and where we have used the relation $2ZecJ_\perp B(s, t) = c^2 p_\perp^2$. We note that the Swann term comes from $\partial B/\partial t$, the partial time derivative of B, while the Fermi term is due to the space derivation $v_B(\partial B/\partial s)$, and they add to give

$$\left(\frac{\partial W}{\partial t}\right)_s + \left(\frac{\partial W}{\partial t}\right)_F = \frac{c^2 p_\perp^2}{2W} \frac{1}{B} \frac{\partial B}{\partial t} .$$

In other words, these two effects combined give the energy change of the particle due to the total time variation of the field.

To lowest order, when gyration is very fast, we may take the average over $\phi_\perp$ (indicated by $\langle\ \rangle_\perp$), and the terms corresponding to the induction effect and the transit-time effect vanish, leaving

$$\left\langle \frac{dW}{dt} \right\rangle_\perp = \left(\frac{\partial W}{\partial t}\right)_S + \left(\frac{\partial W}{\partial t}\right)_F. \tag{IV.29}$$

Similarly, if the longitudinal motion of the particle is sufficiently fast, we may further take the average over pitch-angle variations along the line of force–weighted proportionally to the time Δt a particle of given momentum spends on a line element ds. If we indicate this average by $\langle\ \rangle_\parallel$, we find for the Swann term

$$\left\langle \left(\frac{\partial W}{\partial t}\right)_S \right\rangle_\parallel = \frac{c^2 p^2}{2W} \left\langle \frac{\sin^2 \alpha}{2B} \frac{\partial B}{\partial t} \right\rangle_\parallel, \tag{IV.30}$$

which shows that for this mechanism $\Delta p/\Delta t \propto p$. For the Fermi-term, we obtain

$$\left\langle \left(\frac{\partial W}{\partial t}\right)_F \right\rangle_\parallel = \frac{c^2 p^2}{W} v_B \left\langle \sin \alpha \cos \alpha \frac{\partial \alpha}{\partial s} \right\rangle_\parallel.$$

We are interested in the case where the particle is reflected by a magnetic mirror, changing the pitch angle from an initial value α_1 to α_2, say. It can then be shown that to first order in $v_{B/v}$

$$\left\langle \left(\frac{\partial W}{\partial t}\right)_F \right\rangle_\parallel = \frac{W}{\Delta t} \frac{2v_B v}{c^2} \cos \alpha, \tag{IV.31}$$

which shows that for this so-called Fermi II acceleration $(\partial W/\partial t) \propto W$. Finally, Equation (IV.31) can be averaged with respect to the initial pitch angle α_1, resulting in

$$\left\langle \left(\frac{\partial W}{\partial t}\right)_F \right\rangle_{\parallel,\alpha_1} = \frac{2W}{\Delta t} \left(\frac{v_B}{c}\right)^2. \tag{IV.32}$$

Equation (IV.32) gives the well-known result that the energy gain is proportional to energy, or $W \propto W_0\, e^t$. We may combine the informations on the Swann and Fermi mechanisms contained in Equations (IV.29), (IV.30), and (IV.32) by stating that the main energy variation with time contains two terms, one proportional to momentum, the other proportional to energy of the particle, i.e.,

$$\frac{dW}{dt} = c_1 vp + c_2 W, \tag{IV.33}$$

where

$$c_1 = \left\langle \frac{\sin^2 \alpha}{2B} \frac{\partial B}{\partial t} \right\rangle_\parallel$$

and

$$c_2 = \frac{2}{\Delta t} \left(\frac{v_B}{c}\right)^2.$$

Both Dungey (1958) and Parker (1958) have objected to the use of the Swann mechanism in sunspots, the difficulty being the very high conductivity of the solar plasma. A problem with the Fermi mechanism is that the particles in the long run will experience nearly as many decelerations as accelerations. This makes the net rate of acceleration quite small. Parker tried to mitigate this criticism by proposing that the particles are accelerated between shocks which cross each other. Wentzel (1963) elaborated this idea and showed that the shocks need not be strong to accomplish significant acceleration. It seems that a mechanism relying on acceleration in shocks furnishes the most promising avenue toward an understanding of the acceleration of high-energy particles (see also Schatzman, 1963; Severny, 1964).

(c) Sweet's Mechanism

In this mechanism (Sweet, 1958a, b, 1964; Parker, 1963a), the energy for the particles – and also for the whole flare phenomenon – comes from the annihilation of magnetic fields. Sweet's work is not primarily concerned with the acceleration of particles, but was advanced in the search for a mechanism to speed up the annihilation of magnetic fields (see also Dungey, 1953). The basic principle involves two oppositely directed fields that are pushed together in a steady motion, thereby interdiffusing in times small compared to the decay time for resistive diffusion over distances characterizing sunspot configurations. As the two oppositely directed fields are pushed together, the lines of force will reconnect in a neutral plane, or a narrow current sheet. The mechanism leads to an enhanced rate of reconnection when the field lines are pushed together, so that matter – which normally would impede the approach of the field lines – is squeezed out. The mechanism has been studied further by Green (1965) and Priest (1972a). Let l be the characteristic length of the field gradient across the neutral plane. This length may be several orders of magnitude smaller than the characteristic length, L, of the field configuration itself. While the diffusion time for the ambient field is

$$\tau_{\text{diff}} = \frac{4\pi\sigma L^2}{c^2},$$

(IV.34)

the diffusion time of the field in the region of steep gradient is

$$\tau_{\text{sweet}} = \frac{4\pi\sigma l^2}{c^2},$$

i.e.,

$$\tau_{\text{sweet}} = \left(\frac{l}{L}\right)^2 \tau_{\text{diff}}.$$

(IV.35)

Consequently, the reason why Sweet's mechanism works, i.e., why two oppositely directed fields interdiffuse more rapidly than Equation (IV.34) indicates, is that $l \ll L$ if the fields are pressed together enough.

In Sweet's mechanism the nature of the forces pushing the fields together is not specified, but some external motions are assumed. Also, it is not obvious how the step should be taken that leads from field annihilation to genuine particle acceleration. But since magnetic fields are omni-present in active solar regions and other energy sources seem inadequate, the idea to use the magnetic field energy seems sound.

However, even Equation (IV.35) may not describe a sufficiently rapid diffusion, and several attempts have been made to improve the speed of Sweet's mechanism.

In re-examining Sweet's mechanism, Petschek (1964) pointed out that the conversion of magnetic energy to plasma energy is greatly speeded up if one distinguishes between a diffusion and a convection region in the current sheet. Similar arguments have been advocated by Sonnerup (1970) and Yeh and Axford (1970). We shall consider the conditions near the neutral plane in some detail in the two-dimensional case.

Let the magnetic field lines and the material flow be contained in planes parallel to the XZ plane, and let $\partial/\partial y \equiv 0$. Most studies of the Sweet mechanism assume an X-type null point in the magnetic field; let this be at the origin. The fluid is assumed to flow horizontally toward the null point and to be ejected in the vertical direction. Since this motion would imply that the flow takes place across the field lines, there must be an electric field along the y-axis. The electric conductivity is very high, and the magnetic field lines will be convected with the fluid. This region we call the convection region (Petschek, 1964; Sonnerup, 1970), and approximately

$$\mathbf{E} + \mathbf{v} \times \mathbf{B} = 0 . \tag{IV.36}$$

On the other hand, in a small region, characterized by the width δl, very near the X-type null point, the velocity and the magnetic field both will vanish, but the electric field must remain constant. Consequently, in this so-called diffusion region Equation (IV.36) breaks down, and is to be replaced by Ohm's law

$$\mathbf{j} = \sigma(\mathbf{E} + \mathbf{v} \times \mathbf{B}) . \tag{IV.37}$$

Here the convective transport of the field lines becomes insignificant compared to diffusion. Both Petschek and Sonnerup solved for the convection and diffusion region separately and then joined the solutions at the boundary between the regions.

Again we see that the reason why the application of these ideas should lead to a speeding up of the Sweet mechanism, is that the diffusion region is small compared to the characteristic length l of the field gradient, i.e., $\delta l \ll l$.

Also the Petschek version of Sweet's mechanism has its difficulties (Priest, 1972b), and we are far from a complete understanding of field annihilation and the possible resulting particle acceleration.

(d) The 'Current Limitation' Mechanism

The basic problem to solve for any acceleration mechanism is how and where the electric field is established which accelerates the charged particles. It is not obvious that any of the mechanisms hitherto considered can do this in a convincing fashion.

Carlqvist (1969) and Syrovatsky (1970) have shown that strong electric fields may be generated as the result of a sudden interruption of the electric currents in an appropriate configuration in the solar atmosphere. The mechanism is closely linked to the formation of the flash phase of flares, and could account for particle acceleration during this phase. Carlqvist considered the current break due to instability in an isolated force-free twisted magnetic flux tube (Alfvén and Carlqvist, 1967) (see also Gold and Hoyle, 1960). The instability leads to a space charge separation and a large electric field is established. Syrovatsky, doubting the applicability of such flux tubes in this case, builds his theory on a current sheet. The strong current near the X-type neutral point in the magnetic field is interrupted by a plasma instability (see also Giovanelli, 1946, 1947, 1948; Dungey, 1958; Sweet, 1958a, b; Petschek, 1964; and Syrovatsky, 1966, 1969). In contrast to the steady conditions in Sweet's model, Syrovatsky supposes unsteady motions near the neutral line. As the X-type neutral point closes up, the current $j = n_e v_{eff}$ increases, and as $v_{eff} \rightarrow c$, the plasma instability occurs, leading to a strong electric field.

The interruption of the electric current leads in Carlqvist's case to an electric field of electrostatic origin. In Syrovatsky's model the field is of electromagnetic origin, and is much stronger than the stationary field normally found near the neutral line (see Equation (IV.37).

Objections can be raised against both Carlqvist's and Syrovatsky's work (Smith and Priest, 1972). In the former case instabilities will likely lead to turbulence and not to significant space charge separations. As a result the particles will be heated, not accelerated. Also in the latter case turbulence is likely to play havoc with the particles. If turbulence develops, v_{eff} will not reach the velocity of light, and, as a consequence, the particles diffuse across the field lines and stop the compression before the plasma instability occurs. It seems that particle acceleration in the solar atmosphere remains a major problem.

4.2.3. JEFFERIES AND ORRALL'S THEORY FOR LOOPS

The basic idea in this work (Jefferies and Orrall, 1965) is that the mass of loop prominences is fed into the system in the form of energetic protons at the bottom of the loops. The particles follow the magnetic lines of force until they give up their ordered motion by Coulomb collisions with the ambient gas. This thermalization process is supposed to take place mainly near the top of the loops. The energy thereby released creates a dense hot region, and as the density increases, more and more energetic particles can be trapped. Ultimately, this hot region becomes so dense that it explodes, and matter will stream out along the magnetic lines of force and flow down the two legs of the loops. The expanding plasma will cool enough to be visible in Hα and other optical radiations.

Jefferies and Orrall's model has a number of attractive features, and it also circumvents two difficulties that a condensation process for loops will have to contend with, namely (a) mass requirements and (b) motions across magnetic lines of force. Let us briefly examine these problems before discussing the injection theory.

(a) *Mass Requirements*

In principle, one can determine the density, n_e, and the velocity, v_p, of the prominence material in the loops, as well as the diameter of the loops, and thereby estimate the total mass flux to the photosphere or chromosphere. Jefferies and Orrall assumed $n_e = 10^{11}$ cm^{-3}, $v_p = 100$ km s^{-1}, and arrived at a value in excess of 10^{16} g for the total mass flow, M, during the lifetime of the loops. Similar estimates have been made by Kleczek (1964). Now, loop prominence systems occur in coronal condensations and according to Waldmeier and Müller's (1950) model of such condensations, their total mass is of the order of 10^{15} g. From this, Jefferies and Orrall argued that there is not enough matter available for large loops to condense from the corona, and they proposed that an injection mechanism must be responsible. From this picture, loop prominences are seen to be not condensations but injections, and one may ask if other types that generally give the impression of condensing might in reality originate from below. Whatever the answer, one should remember that those types that seem to originate from above do so in the sense that material first becomes *visible* in the corona, whence it streams down. Hence, in theories implying injection from below, we find differences in the physical state of the injected plasma. For instance, in surges, the material is at a temperature and in a degree of ionization such that it emits in the visible part of the spectrum (as a not-too-hot gas) on its way up; in other words, the plasma is thermalized during its entire lifetime in the chromosphere. As we shall see shortly, this is not so for loops in Jefferies and Orrall's model.

(b) Motions Across Magnetic Lines of Force

There is reason to believe that matter – in addition to coronal condensations, is available to partake in the formation of loop prominences (see Sections 4.6 and 4.7). But even if this additional mass is enough to account for the total mass required, $M \approx 10^{16}$ g, Jefferies and Orrall argue that the difficulty of transporting the plasma across the magnetic field lines still remains. It is, however, hardly possible to evaluate this objection before we gain further insight into the mechanism of certain plasma instabilities that might change the picture significantly. We mention for instance the hydromagnetic instability (Bishop, 1966; Jeffrey and Taniuti, 1966), which accomplishes an interchange in position between plasma and magnetic field, and the (microscopic) effect of particle drift across a strong magnetic field (Spitzer, 1956). If a dense coronal condensation is adjacent to a magnetic flux tube, it is conceivable that either instability might lead to the formation of a dense knot of plasma in the field and that the loop might grow out of this knot.

Returning now to Jefferies and Orrall's injection model, we notice that as the particles become thermalized near the top of the loop, they produce heat, much of which will be lost by radiation. The authors made an estimate of the net radiant energy loss by using an expression for an optically thin solar plasma derived by Orrall and Zirker (1961),

$$L_{rad} \approx 2 \times 10^{-23} n_e^2 . \tag{IV.38}$$

The total energy loss from the whole loop system due to radiation is then

$$L_{rad, tot} = \frac{L_{rad}}{n_e} \frac{M}{m_H} t_{loop} , \tag{IV.39}$$

where t_{loop} is the lifetime of an individual loop in the system. Also, $t_{loop} \approx 10^3$ s and $M/m_H = 6 \times 10^{39}$ cm^{-3} for a total mass $M = 10^{16}$ g. Equations (IV.38) and (IV.39) give

$$L_{rad, tot} = 2 \times 10^{20} n_e .$$

The electron density is difficult to estimate. It is probably greater than 10^{11} cm^{-3}, which means that the total radiative loss is several times 10^{31} erg. The model implies that there is energy balance between the radiative loss and the kinetic energy, G_{kin}, delivered to the loop by the fast particles. If all the particles have the same initial velocity, v, the total kinetic energy is $G_{kin} = \frac{1}{2} M v^2$ and

$$v^2 = 20 n_e t_{loop} \, .$$

Jefferies and Orrall assumed $n_e = 5 \times 10^{11}$ cm^{-3} and concluded that the energy balance of this type of active object can be maintained by protons with velocities 10^8 cm s^{-1}, or energy about 10^4 eV.

To get these particles up into the corona to form loops they can hardly be accelerated at photospheric levels. In their treatment of the injection mechanism Jefferies and Orrall relied on the concept of particle storage in coronal magnetic fields. The particles are assumed to be generated during those flares that always seem to precede loops, but no theory for this process is available.

As the cooled prominence material flows down into the photosphere, it emits a spectrum, the lines of which should reveal characteristic profiles. For instance, the wings of Hα should be quite pronounced and be different from Stark-broadened lines. It is known that such profiles are actually observed in loops.

Jefferies and Orrall drew attention to the possibility that other active prominences, and even quiescent objects, may be produced by the injection mechanism. If the thermalization of the injected particles takes place near the top of the loop of the magnetic field, a loop prominence should be formed as described. But if it occurred near the bottom of one side of a loop-formed flux tube, the result might be surge-like object or a loop in which matter is seen to stream up one of the legs and down the other.

4.2.4. DIAMAGNETIC EFFECTS

(a) The 'Melon-Seed Effect'

Severny and Khoklova (1953) and Schlüter (1957b) have suggested an interesting mechanism by which an aggregate of charged particles, a plasma cloud, may be accelerated in a magnetic field. It is assumed that the matter to be injected is not magnetized and therefore moves easily between magnetic lines of force. The plasma cloud behaves as a diamagnetic body, and it is the Maxwell tension of the magnetic field that accelerates it. The mechanism is referred to as the melon-seed effect, since the plasma is being squeezed out between the field lines like a melon seed between two fingers. It is made possible by the high electric conductivity of the plasma cloud. The force acting on the clouds is given by $\nabla \ln B^2$, where $\mathbf{B}$ is the magnetic field in the absence of the unmagnetized clouds.

If effective, the mechanism may provide a means of explaining the surge-like ejections observed at times of flare. The outward acceleration is given by $dv/dt = -\nabla[kT \ln B^2 + \Phi]$, where Φ is the gravitational potential.

(b) Jensen's Injection Mechanism

Jensen (1959) has shown that when a plasma is not in thermal equilibrium, an inhomogeneous magnetic field will cause the plasma to move in the field. Under certain conditions the plasma can be forced to move up into loop structures in the field, thereby conceivably accounting for some loop prominences and coronal condensations. From the Vlasov equations for either the ion or electron distribution functions f (which relate to the microscopic velocities $\mathbf{w}$)

$$\frac{\partial f}{\partial t}+w_\parallel\frac{\partial f}{\partial z}-\frac{w_\perp^2}{2B}\frac{\partial B}{\partial z}\frac{\partial f}{\partial w_\parallel}-g\frac{\partial f}{\partial w_\parallel}+\frac{w_\parallel w_\perp}{2B}\frac{\partial B}{\partial z}\frac{\partial f}{\partial w_\perp}=0,$$

where we have introduced

$$\frac{\partial w_r}{\partial t}=0,\quad\frac{\partial w_\perp}{\partial t}=\frac{w_\parallel w_\perp}{2B}\frac{\partial B}{\partial z},\quad\frac{\partial w_\parallel}{\partial t}=-\frac{w_\perp^2}{2B}\frac{\partial B}{\partial z}-g\,.$$

Jensen derived a linearized equation of motion in $v_\parallel$, the macroscopic velocity parallel to the magnetic field

$$\rho\frac{\partial v_\parallel}{\partial t}+\frac{\partial p_\parallel}{z}+\frac{n}{B}\frac{\partial B}{\partial z}(W_\perp-2W_\parallel)+\rho g=0,\qquad\text{(IV.40)}$$

where

$$\mathbf{v}(\mathbf{r},t)=\frac{1}{n(\mathbf{r},t)}\int_{-\infty}^{\infty}\int_{-\infty}^{\infty}\int_{-\infty}^{\infty}\mathbf{w}f(\mathbf{r},\mathbf{w},t)\,dw_x\,dw_y\,dw_z\,.$$

$W_\perp$ and $W_\parallel$ are the kinetic energies associated with motion perpendicular and parallel to the magnetic field, respectively, and the field was assumed to have the following components in cylindrical coordinates

$$\mathbf{B}=\left[-\frac{1}{2}r\frac{\partial B}{\partial z},0,B_z\right].$$

The pressure $p_\parallel$ relates to the component of the random velocity $u_\parallel=w_\parallel-v_\parallel$ by the assumption $\partial(nmu^2/2)/\partial z=\partial p_\parallel/\partial z$. Equation (IV.40) shows that if we have thermal equilibrium ($W_\perp=2W_\parallel$, since there are two degrees of freedom in the perpendicular direction, but only one in the parallel direction), the magnetic field has no influence on the density distribution (van de Hulst, 1950). But if there are deviations from equipartition, an inhomogeneous ($\partial B/\partial z\neq0$) magnetic field will change the density distribution of the plasma. This result was derived using the Vlasov equation, which means that the effect of collisions cannot be assessed. With $\partial B/\partial z=-|\partial B/\partial z|$, the hydrostatic equation becomes

$$\frac{\partial p_\parallel}{\partial z}=\frac{n}{B}\left|\frac{\partial B}{\partial z}\right|(W_\perp-2W_\parallel)-\rho g\,.\qquad\text{(IV.41)}$$

Provided one can define a temperature $T_\parallel$ that is independent of height, Equation (IV.41) can be integrated to give the density distribution as a function of height

$$n(z)=n_0\exp\left\{\int_0^z\left[\frac{W_\perp-2W_\parallel}{2W_\parallel B}\left|\frac{\partial B}{\partial z}\right|-\frac{mg}{kT_\parallel}\right]dz\right\}.$$

Jensen assumed that the deviations from equipartition ($W_\perp\neq2W_\parallel$) were caused by a varying magnetic field, since only $W_\perp$, and not $W_\parallel$, is altered by induction. If $\partial B/\partial t>0$, $W_\perp$ is increased and $W_\perp-2W_\parallel>0$, which means that the plasma is diamagnetic and it will be pushed to regions where the magnetic field strength has a minimum. (The condition $\partial B/\partial t<0$ leads to a paramagnetic plasma which moves to places where the magnetic field is strongest (see Kiepenheuer, 1938). To maintain a positive gradient of density against gravity, the ratio of scale-height, $H = kT_{\parallel}/mg$, to characteristic length, L, of the magnetic field must satisfy the condition

$$H/L > 2W_{\parallel}/(W_{\perp} - 2W_{\parallel}) . \tag{IV.42}$$

One can find the time-scale for the variation in the magnetic field necessary to maintain deviations from thermal equilibrium large enough to be of interest by writing for the variation of the ion component of $W_{\perp} - 2W_{\parallel}$ with time (see Schlüter, 1957a):

$$\frac{\partial(W_{i, \perp} - 2W_{i, \parallel})}{\partial t} = -\frac{W_{i, \perp} - 2W_{i, \parallel}}{\tau_i} + \frac{W_{i, \perp}}{B} \frac{\partial B}{\partial t} ,$$

where τ_i is the mean time between collisions for ions. Any deviation from an isotropic distribution of velocities will be smoothed out in time intervals of the order of τ_i. We require that $\partial(W_{i, \perp} - 2W_{i, \parallel})/\partial t = 0$ to maintain the difference $W_{i, \perp} - 2W_{i, \parallel}$. Then, defining the time-scale for the variation of the magnetic field $t_B = |(1/B)(\partial B/\partial t)|^{-1}$, we find

$$t_B = \frac{W_{i, \perp}}{W_{i, \perp} - 2W_{i, \parallel}} \tau_i . \tag{IV.43}$$

For the electrons we may assume thermal equilibrium, $W_{e, \perp} = 2W_{e, \parallel}$, since the relaxation time for the electrons is much shorter than for the ions. This gives, with Equations (IV.42) and (IV.43), the condition for the density gradient

$$t_B < \frac{W_{i, \perp}}{2(W_{i, \parallel} + W_{e, \parallel})} \frac{H}{L} \tau_1 . \tag{IV.44}$$

For example, for a scale height ten times greater than the characteristic length, we find from Equation (IV.42) that $W_{\perp} > 2 \cdot 2W_{\parallel}$, and if τ_i is of the order of 10 to 100 s, Equation (IV.44) requires t_B to be less than a few minutes.

Consequently, dense regions can be formed as a result of this diamagnetic effect if the magnetic field undergoes changes on a time-scale of the order of a few minutes. Such rapid changes may not be unreasonable in certain active regions.

4.2.5. SIPHON-TYPE INJECTIONS

In addition to their role in Jensen's mechanism, magnetic flux tubes have been invoked in another way to transport matter from photospheric layers up into the corona to form prominences. The principle on which this siphon-like action works is due to the fact that the plasma in the flux tube cannot be in equilibrium if conditions (pressure or temperature) differ at the ends of the arched flux tube (assumed to be at the same height). A material flow will then take place along the arch.

(a) Meyer and Schmidt's Magnetically Aligned Flow

In an attempt to account for the Evershed effect in sunspots, Meyer and Schmidt (1968) studied the hydrostatic equilibrium conditions in a magnetic arch whose feet in the photosphere were in regions of different gas pressure. This pressure difference will drive a quasi-stationary flow along the flux tube. Meyer and Schmidt showed that the flow velocity reaches the sound speed at the top of the arch, and this condition determines the amount of mass that can be transported per second for a given density. Furthermore, the authors claim that due to gravity the density decreases vary nearly exponentially as a function of the maximum height of the arch. If this is true, the total mass flow also decreases exponentially with the maximum height of the arch. Hence, the mass transport from the photosphere is significant only for relatively low-lying flux tubes. Meyer and Schmidt found that the fluid may reach supersonic velocities after it has passed the top of the arch and is on its way down. Finally, it adjusts to the pressure further down by a shock front.

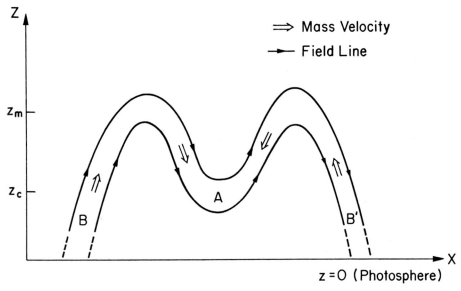

Fig. IV.4. The magnetic field configuration and material motion in Pikel'ner's (1971) model, shown schematically.

(b) *Pikel'ner's Siphon Mechanism for Quiescent Prominences*

In order to bring material up into coronal regions where quiescent prominences are found, Pikel'ner (1971) considered the field configuration depicted in Figure IV.4. (Similar arches, placed parallel to each other in the y-direction, provide the necessary 3-dimensional arrangement.) Essential to his theory is the pronounced bending down of the top of the arches, whereby a 'pit' is formed at A. Pikel'ner argues that the heating of the gas at A is small. The main energy is transported to the coronal regions by Alfvén waves and slow waves (Pikel'ner and Livshitz, 1964), and the magnetic field shields the gas at A from the heating from below. The energy flux, F, that is put into the tube from chromospheric or photospheric layers (at B and B'), is conducted along the tube toward A, but is gradually exhausted on its way due to radiative cooling, L_{rad}. Because of this there will be a temperature gradient along the tube, gas at A being at a significantly lower temperature than at either B or B'. Hence, as in Meyer and Schmidt's (1968) model, material will start to flow along the flux tube toward A as indicated by the arrows in Figure IV.4. Pikel'ner argues that instead of supersonic flow, as found by Meyer and Schmidt, his model probably calls for subsonic velocities, since the scale height in the corona and the height of the arches are similar.

With the temperature decreasing upward along the tube, the thermal conductivity decreases and eventually the material, near A, becomes a dense, cool gas, which Pikel'ner identifies with part of a quiescent prominence. This condensation is not due to the thermal instability considered previously, but is the result of the (macroscopically) decreased heating.

Pikel'ner assumes the following form for the shape of the tube from B (or from B') to A:

$$\frac{z}{z_m} = 1 - \left(\frac{x}{x_0}\right)^2 ,$$

where z_m is the maximum height of the arch and $2x_0$ is the horizontal extent of the arch. Flux constancy and cross-sectional area, A, of the tube are related by $A/A_0 = B_0/B$, where B_0 is the magnetic field strength at the feet of the arch. Rewriting, we find

$$A = A_0 f_B(z) ,$$

where the function describing the change of the field with height is taken to be

$$f_B(z) = 1 + (W_m - 1) \frac{z}{z_m} .$$

Pikel'ner took $z_m = 10^5$ km, $x_0 = 7 \times 10^4$ km and $W_m = 10$, and assumed steady flow. The constancy of entropy, s, where $dQ/T = ds$, or, per unit volume,

$$\rho \frac{ds}{dt} = \frac{\rho}{T} \frac{dQ}{dt} , \tag{IV.45}$$

permitted (since $\partial s/\partial t = 0$):

$$\rho T v \frac{ds}{dt} = \rho T v \left[\frac{c_v}{T} \frac{dT}{dl} + c_v(1-\gamma) \frac{1}{n} \frac{dn}{dl} \right] , \tag{IV.46}$$

as an energy equation (instead of Equation (IV.9)) where dl is a line element along the flux tube. In Equation (IV.46) we have made use of the expression $s = s_0 + c_v \ln (p\rho^{-\gamma}/\gamma - 1)$ for the entropy and assumed a perfect gas law. Equations (IV.45) and (IV.46) combine to give the energy equation.

$$\rho c_v v \frac{dT}{dl} + \rho c_v(1-\gamma) \frac{T}{n} v \frac{dn}{dl} + L_{rad} - \nabla \cdot \left(K \frac{dT}{dl} \right) = 0 , \tag{IV.47}$$

since in the heat gain or loss per unit volume, $\rho(dQ/dt)$, Pikel'ner included only radiation losses, L_{rad}, and the effect of thermal conductivity, $\nabla \cdot (K \nabla T)$. Equation (IV.47) was combined with the equation of mass conservation, $nvA = \text{const}$, or

$$nv W(z) = C_1 \tag{IV.48}$$

and the equation of motion (Equation (IV.15)) in a steady state, or

$$v \frac{dv}{dt} + \frac{1}{p} \frac{d\rho}{dl} + g \frac{dz}{dl} = 0 . \tag{IV.49}$$

The system of Equation (IV.47) to (IV.49) was solved numerically with the boundary conditions: at $l=0$, $n_0=2\times10^8$ cm^{-3}, $T_0=1\cdot4\times10^6$ K and all energy dissipated into the tube arises from thermal conduction; any mechanical wave is assumed to have been absorbed lower down.

Values of $C_1\approx10^{14}$ to 10^{15} s^{-1} will give sufficient density increases at A to be considered prominence conditions. The corresponding flow velocities reach values of $v\approx10$ km s^{-1}, but decrease rapidly at $z=z_c$ (see Figure IV.4) in the cool gas.

The height, z_c, where the condensing of the prominence takes place, depends strongly on the conditions at B and B', as well as on the energy flux. By allowing the arch to start in areas with the appropriate value of F, Pikel'ner seems to be able to make the prominence formation take place at reasonable heights. As indicated by the author, improved results may be obtained with a non-stationary model allowing for some flux to be transported by waves, i.e., by adding a gain term, G_{mech}, to Equation (IV.47). However, other boundary conditions should be explored since T_0 does not seem reasonable, and improved expressions for L_{rad} should be used. Pikel'ner put $L_{rad}=n^2f(T)$ where $f(T)$ is the function calculated by Pottasch (1965).

4.2.6. REMARKS ON SURGES AND SPRAYS

While both condensation and injection have been invoked for the formation of most types of prominences (quiescent filaments, loops, coronal rain, etc.), only injection is applicable to surges and sprays. These types of prominence definitely originate from below, and provide the best example of injections.

Most surges occur at the outer edge of sunspot penumbrae, and the correlation with flare occurrence is high (Kiepenheuer, 1968; Rust, 1968). The total mass ejected in a surge ranges between 10^{14} and 10^{15} g. The number density is probably around 10^{12} cm^{-3}; Kiepenheuer gives higher values, assuming surges to be injected at very low, even photospheric, heights. The velocity, v, of surges ranges from about 100 km s^{-1} to several times this value. Sprays have velocities in excess of the escape velocity, $v_{esc}=612$ km s^{-1} and at times exceeding 1000 km s^{-1}. The corresponding kinetic energy density ranges from 10^2 to maybe 10^4 erg cm^{-3}.

The path of surges generally is directed away from the nearest sunspot (Giovanelli and McCabe, 1958; Shaposhnikova, 1958), indicating the influence of strong magnetic fields. There seems to be little doubt that the surges move along magnetic field lines, but it is not *a priori* clear whether the surge moves along pre-existing field lines or drags the field with it out in the corona.

Observations (Section 2.3) show that surges have magnetic fields of strength up to about 150 G. Simply equating the corresponding magnetic energy density, $B^2/8\pi$, and the kinetic energy density, $\frac{1}{2}mv^2$, we find that surges with such fields are magnetically dominated for velocities less than about 300 km s^{-1}. To control an object whose velocity equals v_{esc} would require a magnetic field of at least 275 G. Therefore, while sprays ($v>v_{esc}$) cannot be contained by the magnetic fields, the motion of the slower surges is completely determined by the fields.

Rust (1968) has analyzed the motions of surges and their relation to changes in the magnetic field of sunspot groups. He argues that – at least for some cases – the kinetic and radiative energy dissipated by a surge is comparable to an accompanying decrease in magnetic field energy. This might indicate that annihilation of the magnetic field –

which in Rust's picture takes place near a neutral point over a satellite sunspot – provides the energy for acceleration of chromospheric matter in the form of surges. He further showed that some low-energy surges ($v\sim100$ km s^{-1}) follow magnetic lines of force as constructed from potential field theory (see Section 3.2.3).

We conclude that surges are injected – by a yet to be determined acceleration mechanism that derives its energy from sunspot magnetic fields – along certain flux tubes above active regions. For sprays the question of their path in relation to the field configuration is academic; the dynamics of such objects is not determined by the existing fields. However, also in these cases the acceleration of the prominence material probably is due to the magnetic field.

4.3. Spicules

In many respects spicules may be considered 'mini-surges' and treated as prominences. Yet spicules are an integral part of the physics of the quiet Sun, and their discussion belongs in a treatise on the chromosphere. Either point of view could be followed; we shall adhere to the former.

4.3.1. INTRODUCTION

Roughly speaking, spicules are cylindrical or cone-shaped objects with a diameter of about 1000 km reaching from the low chromosphere 6000 to 10 000 km into the corona. Their lifetime is about 5 to 10 min and the whole spicule shows an upward motion of 20 to 30 km s^{-1}. After their ascent many spicules diffuse and fade out, while others are seen to descend, seemingly following in reverse the path of ascent. For a detailed discussion with references to spicule observations see Beckers (1968a).

Not only the diameter and lifetime but also the birth rate of spicules are similar to those of the photospheric granulation. At any one time there may be about 10^5 spicules present, or a few times this value. Consequently, the granulation may play a role in the formation of spicules (Beckers, 1966).

Studies of emission-line profiles from spicule spectra indicate that the electron temperature is of the order of $1{\cdot}5\times10^4$ K, and the electron density about 1 to 2×10^{11} cm^{-3}, much cooler and denser than the interspicular region, as found for surges and other prominences. The values quoted are averages pertaining to heights around 4000 to 6000 km, but line-width measurements indicate that n_e and T_e vary both across a spicule and with height. The metal-line emission comes from the cooler regions, He lines from the hotter parts of the spicule plasma.

4.3.2. THE ROLE OF SUPERGRANULATION AND MAGNETIC FIELDS

Spectroheliograms of the solar disk taken in strong lines show a bright pattern, called the chromospheric network. This network coincides with the boundaries of supergranulation cells (Simon and Leighton, 1964), and the boundaries lie in regions of enhanced magnetic field in the photosphere. Although the photospheric field is generally weak in other parts of the quiet Sun (a few gauss), it exceeds 30 G along boundaries of supergranulation cells. Since gas motion in the cells is (horizontally) from the center toward the boundaries, it is believed that the enhanced magnetic fields are caused by the supergranulation (Parker, 1963b).

It is along these boundaries that we find the spicules. Therefore, it looks as if the formation of spicules is intimately connected with enhanced magnetic fields in the photosphere, and that in turn the magnetic fields owe their enhancement to the super-granulation. If this picture is correct, the spicules have at their disposal at least part of the vast energy reservoir of the subphotospheric convection. In this way the spicule phenomenon is much more systematically tied in with the structure of the quiet Sun than all the prominences hitherto considered.

4.3.3. ORIGIN OF SPICULES

From the observations cited above we may picture the spicules as those regions of the chromosphere through which the strongest magnetic fields are channeled. The field lines run more or less along the axis of the spicules and prevent them from dispersing quickly into the coronal environment. Equality of the kinetic energy density of the spicular plasma, $\frac{3}{2}nkT$, and the magnetic energy density requires, for $T_e = 1 \cdot 5 \times 10^4$ K and $n_e = 2 \times 10^{11}$ cm^{-3}, a magnetic induction of only $B \approx 5$ G.

Osterbrock (1961) considered spicules the manifestation of hydro-magnetic waves carrying energy up into the corona from subphotospheric layers. In this picture the spicules *are* the chromosphere, constituting the transition region between the photo-sphere and corona and providing the coupling between the two. Osterbrock discussed in some detail the enhanced wave-generation possible in plage regions, and the idea has been further developed by Parker (1964) and Kuperus (1965). Basically, it rests on the fact that the mechanical energy flux, $\mathbf{F}_{mech}$, increases when the vertical magnetic field increases.

In Parker's model magneto-acoustic waves develop into shocks as they travel up into the atmosphere, and the shocks are identified with the spicules. However, no detailed calculations explain how this shock-to-spicule mechanism takes place. Parker's work has been developed further by Wentzel and Solinger (1967). They started out with an initial non-thermal photospheric motion of $1 \cdot 5$ km s^{-1} and studied the development of the shock. With a magnetic field of 25 G they showed that the resulting gas behind the shock would be heated to between 6000 K and 31 000 K, depending on whether the gas was isothermal ($\gamma = 1$) or adiabatic ($\gamma = \frac{5}{3}$). The density for these two extreme cases came out to be 2×10^{10} cm^3 and 4×10^{10} cm^{-3} respectively, which, in either case, is too low.

The shock-wave model has many attractive features, tying, as it does, the spicule formation to the important characteristics of waves in the solar atmosphere. However, there are several unanswered problems and it is not obvious how the spicules, as a dissipating shock, will fall back again into the photosphere.

Kuperus and Athay (1967) used the dissipation of the mechanical flux, $\mathbf{F}_{mech}$, to heat the chromosphere. The corresponding term in the energy equation, Equation (IV.10), $G_{mech} = \nabla \cdot \mathbf{F}_{mech}$, is balanced by radiation and conduction losses, L_{rad} and L_{cond} $(= \nabla \cdot [K \nabla T])$. Since they inlcude L_{cond}, instead of Equation (IV.20), Kuperus and Athay used, in a steady state,

$$c_p \frac{\partial T}{\partial t} = G_{mech} - L_{rad} - \nabla \cdot (K \nabla T) = 0 . \tag{IV.50}$$

In the expression for G_{mech}, Equation (IV.13), they took $C_1 = K_1 V_s^2 \phi(M)$, where K_1

is a constant, V_s the speed of sound and $\phi(M)$ a slowly varying function of the Mach number M (Kuperus, 1965). The radiation losses, Equation (IV.14), they approximated by $L_{rad} = \rho^2 f(T)$, and used for $f(T)$ Athay's (1966) values.

It turns out that Equation (IV.50) leads to unacceptably large conduction losses in a shallow layer at the top of the chromosphere. Kuperus and Athay remedied this by including energy losses due to acceleration of matter. With this kinetic energy loss, Equation (IV.50) takes the form

$$G_{mech} - L_{rad} - L_{kin} - \nabla \cdot (K \nabla T) = 0 , \qquad (IV.51)$$

and Kuperus and Athay argued that the shallow layer in the chromosphere which becomes unstable is forced to a dynamic behavior. They identify this dynamic configuration, in which matter is accelerated with spicule formation. It is interesting that this approach, which differs radically from the shock-wave picture discussed above, and which considers the spicules as jets caused by an instability of the Rayleigh-Taylor type in the chromosphere, gives reasonable predictions for the spicule velocities.

The kinetic energy losses correspond to the kinetic energy flux in spicules, $L_{kin} = \nabla \cdot \mathbf{F}_{sp}$, where

$$F_{sp} = \tfrac{1}{2} \rho v_{sp}^2 A v_{sp} .$$

Taking for the fractional area of the Sun covered by spicules, $A = 0.1$, and assuming a density corresponding to $n = 2 \times 10^{11}$ cm^{-3}, and a flux $F_{sp} = 3 \times 10^5$ erg cm^{-2} s^{-1}, we find with Kuperus and Athay a spicule velocity of $v_{sp} = 20$ km s^{-1}. Several other models for accelerating spicules have been proposed, but a comprehensive review of such models would be a digression from the main theme of this book.

The existing models reveal some of the basic physics that may be at work in spicules, even though none of them provides an easy explanation for all the observed characteristics.

CHAPTER V

STABILITY OF PROMINENCES

In discussing models of prominences, and the way these objects form, we have repeatedly touched on the problem of their stability. In this chapter we shall discuss in more detail different equilibria for prominences, and see how stable they are against various perturbations.

For quiescent prominences static conditions are nearly always implied, and often seem to furnish a reasonable approximation. This is not so for active prominences. Here either more complicated equilibria are maintained, and no thorough investigation of their stability is available, or the active object never reaches any degree of equilibrium during their short lifetime.

5.1. Thermal Equilibrium – The Heating Problem

Since quiescent prominences remain close to stationary for long periods of time, we conclude that all the energy flowing into the prominence ($G = G_{rad} + G_{cond}$, see Section 4.1) must be transported out of it at the same rate ($L = L_{rad}$). Poland and Anzer (1971) have discussed the conditions resulting from a study of the equation $G - L = 0$; we shall return to their investigation below.

5.1.1. CONDUCTIVE HEATING

If the radiation losses were too small, the prominence would quickly heat up. We can find a lower limit to the time it would take to heat the plasma from an initial temperature T_p to coronal values, T_c, due to thermal conduction alone, by neglecting radiative losses, i.e., by writing Equation (IV.10) in the form (Rosseland *et al.*, 1958; Ioshpa, 1965)

$$c_v \frac{\partial T}{\partial t} = \nabla(K\nabla T).$$ (V.1)

Following Ioshpa, we consider isobaric heating, put $K = AT^{5/2}$ and obtain in the one-dimensional case the equation

$$\frac{\partial T}{\partial t} = \frac{2A}{5kn_e} \frac{\partial}{\partial x} \left(T^{5/2} \frac{\partial T}{\partial x} \right),$$ (V.2)

which yields the approximate solution

$$T = T_c \left[1 - \frac{x}{(a + T_c^{5/2})^{1/2}} \right]^{2/5},$$ (V.3)

subject to the boundary conditions $T(0, t) = T_c$, $T(x, 0) = 0$.

Equation (V.3) represents a thermal wave with a steep front. At any time t, the width of the zone in which the temperature drops from $0.4T$ to 0 is approximately $1/10$ the thickness of the already heated region. If, for example, the prominence has a thickness of 10^9 cm, it will be heated in $t = 2.5 \times 10^3$ s, and if the prominence consists of fine structure threads of thickness 10^8 cm, the time to heat it to coronal temperatures is 25 s. Similar results are due to Severny and Khokhlova (1953) and Shklovsky (1965). For a more extensive discussion of thermal waves see Zel'dovich and Raizer (1967).

Since quiescent prominences reveal nearly static conditions, these estimates indicate the efficiency with which radiation losses strike a balance. In addition, the existence of a magnetic field in the prominence with a significant component transverse to the heat flow will drastically alter these estimates. While heat conduction along the field is the same as in the absence of the field, the conduction perpendicular to the field is reduced by the factor $1 + (\omega_B \tau)^2$ where $\omega_B = eB/mc$ is the cyclotron frequency. Therefore, to increase the time-scale above by a factor of 10^4, for example, it is sufficient to have a transverse magnetic field of less than 1 G. However, even though the transverse conductivity is greatly reduced, there will be an additional heat flow in a direction perpendicular to both the magnetic field and the temperature gradient, i.e., in the direction $\mathbf{B} \times \nabla T$ (the Righi-Leduc effect). The corresponding conductivity is $\omega_B \tau$ times the direct conductivity and may be of importance for some geometrical configurations in prominences (Tandberg-Hanssen, 1960; Orrall and Zirker, 1961).

5.1.2. THE BALANCE $L_{rad} - G_{rad} - G_{cond} = 0$.

In their work Poland and Anzer (1971) analyzed the conditions when the net radiative losses $(L_{rad} - G_{rad})$ balance the heat gained by conduction (G_{cond}). Hydrogen is the dominant element in prominences, and since H lines dominate prominence spectra, Poland and Anzer assumed that the radiative energy transfer in a quiescent prominence is determined by H. They solved the transfer problem using the method described in Section 3.1.1 (see Equation (III.14)) and showed that the main radiative energy gain is due to the Lyman continuum, and the main loss to Hα. Other sources, such as free-free emission, the He I Lyman continuum Ly$_c$, and probably also the Ca II, K and H lines, are negligible in comparison.

The energy gain in Ly$_c$ can be found from (Thomas and Athay, 1961)

$$G_{rad} \approx \varepsilon(Ly_c) = 2\pi \int_{\nu_0}^{\infty} d\nu \int_{0}^{\tau_{\nu,max}} \rho S_\nu \, d\tau_\nu = \frac{2\pi k T_e}{h} \int_{0}^{\tau_{0,max}} \rho S_0 \, d\tau_0 , \qquad (V.4)$$

where the net radiative bracket $\rho \equiv 1 - (\int J_\nu \phi_\nu \, d\nu / \int S_\nu \, d\nu)$, τ_0 is the optical depth at the head of the continuum, $\tau_{0,max}$ the corresponding maximum optical depth in the prominence, and $\tau_{\nu,max}$ the maximum optical depth reached in the prominence at frequency ν. Poland and Anzer computed the energy loss in Hα by assuming that all collisions from the $n = 2$ to the $n = 3$ level result in the emission of an Hα photon, and that Hα is effectively thin ($\tau_{0,max} < 20$), so that all photons created will escape. Under these conditions the energy loss in Hα on one side of the model prominence sheet of thickness X will be

$$L_{rad} \approx \varepsilon(H\alpha) = \overline{N}_2 \overline{C}_{23} \frac{h\nu X}{2} , \qquad (V.5)$$

where $\bar{N}_2$ is the average $n=2$ population and $\bar{C}_{23}$ is the average collisional rate between levels 2 and 3.

For a model prominence with $T_e = 6000$ K, $n(H) = 10^{11}$ cm^{-3} and $X = 6000$ km; Poland and Anzer found

$$G_{rad} = 9 \times 10^4 \, \text{erg cm}^{-2}\,\text{s}^{-1}, \quad L_{rad} = 2 \times 10^5 \, \text{erg cm}^{-2}\,\text{s}^{-1},$$

so that the net radiation loss from the prominence is about 1×10^5 erg cm^{-2} s^{-1}.

For the heating of the prominence plasma due to conduction the authors used the expression

$$G_{cond} = K T^{5/2} \frac{dT}{dx}, \tag{V.6}$$

where $K = 8\cdot8 \times 10^{-7}$ (Oster and Sofia, 1966), and where they assumed no heat sink in the corona, so that all the heat is conducted into the prominence. With the boundary conditions $T = T_p$ at $x = \frac{1}{2}X$, and $T = T_c$ at $x = \frac{1}{2}X + D$, Equation (V.6) integrates to give

$$G_{cond} = \frac{2}{7} \frac{T_c^{7/2}}{D} K, \tag{V.7}$$

where a term $T_p^{7/2}$ has been neglected in comparison to $T_c^{7/2}$.

Poland and Anzer included the effect of a homogeneous magnetic field, making an angle γ with the prominence sheet. The distance D, over which the temperature drops from coronal to prominence values, must then be replaced by $D/\sin \gamma$, since the heat flows along the field lines. Then Equation (V.7) takes the form

$$G_{cond} = \frac{2}{7} \frac{T_c^{7/2}}{D} K \sin \gamma. \tag{V.7'}$$

Equation (V.7') shows that for coronal temperatures in excess of 5×10^5 K the conductive gain will equal or exceed the radiation loss (1×10^5 erg cm^{-2} s^{-1}) for all values of $D < 500$ km. For a million degree corona, D needs only be smaller than a few thousand km to provide the necessary conductive energy flow to balance the radiative loss.

From this analysis Poland and Anzer concluded that an energy balance is possible in quiescent prominences. However, they ran into the difficulty that the balance requires a H density $n(H) \approx 3 \times 10^{12}$ cm^{-3}. To support the corresponding great mass by Lorentz forces requires a strong magnetic field, probably of the order of 20 G, which is only rarely found in quiescent prominences.

The way out of this dilemma is not quite clear, but Poland and Anzer advocate the possibility that the temperature is greater than 6000 K in prominences. For example, if $T_p = 7500$ K, a density of $n(H) = 5 \times 10^{11}$ cm^{-3} would suffice, and the requirement as to the magnetic field strength would be lessened correspondingly.

5.1.3. GRAVITATIONAL ENERGY RELEASE

So far we have not considered explicitly the influence of gravitational potential energy on the energy balance. A prominence plasma supported by magnetic fields may be subjected to gravitational instabilities (Coppi, 1964), and Sturrock and Coppi (1964, 1965) have invoked such instabilities in their flare model. Even though this discussion leads us to the subject of dynamic equilibrium, which will be treated in Section 5.2, we shall include the discussion here to be able to consider some influences on the heat balance.

Let us follow Coppi's treatment and consider a plasma layer threaded by a horizontal magnetic field under gravity. The equilibrium magnetic field is assumed to be properly sheared, so that the system is stable in the absence of dissipation. Furthermore, let all equilibrium quantities be functions of the vertical coordinate, z, only. The equations governing the problem are the hydrodynamic equations, including the Lorentz force, i.e., Equations (IV.1), (IV.9) and (IV.15), and the electromagnetic Equations (IV.16) to (IV.19).

In the energy equation, Equation (IV.9), we include thermal conduction and Joule heating in the term Q. This leads to the following set of equations:

$$\frac{\partial \rho}{\partial t} + \nabla \cdot (\rho \mathbf{v}) = 0 \tag{V.8}$$

$$\rho \frac{d\mathbf{v}}{dt} + \nabla p - \rho g - \frac{1}{c} \mathbf{j} \times \mathbf{B} = 0 \tag{V.9}$$

$$\frac{dp}{dt} - \gamma \frac{p}{\rho} \frac{d\rho}{dt} - (\gamma - 1) \left[\nabla \cdot (K \nabla T) + \frac{j^2}{\sigma} \right] = 0 \tag{V.10}$$

$$\nabla \times \mathbf{B} - \frac{4\pi}{c} \mathbf{j} = 0 \tag{V.11}$$

$$\frac{1}{c} \frac{\partial \mathbf{B}}{\partial t} - \nabla \times \mathbf{E} = 0 \tag{V.12}$$

$$\nabla \cdot \mathbf{B} = 0 \tag{V.13}$$

$$\mathbf{j} - \sigma \left(\mathbf{E} + \frac{1}{c} \mathbf{v} \times \mathbf{B} \right) = 0 . \tag{V.14}$$

Coppi showed that the only equilibrium condition for the described model subject to the set of Equations (V.8) to (V.14) is of the form

$$\mathbf{B} \cdot \frac{d\mathbf{B}}{dz} = -\rho g - T \frac{d\rho}{dz} .$$

If the system is perturbed about the equilibrium, instabilities may develop. Let the displacement brought about by the perturbation be d and consider solutions of the form $d(r, t) = d(z) \exp [i(\omega t - \mathbf{k} \cdot \mathbf{r})]$ where $\mathbf{k}$ is the wave vector. The electric field, $E_{\parallel}$, parallel to the lines of force of the magnetic field will tend to decouple the plasma motions and the magnetic field lines, thereby opposing the electric field, $E_{\perp}$, perpendicular to the magnetic field, which tends to prevent motions of the plasma across the field lines. Then, unless the magnetic field is sufficiently sheared, the system will be unstable against gravitational modes. Coppi found that when the magnetic pressure is much greater than the gas pressure, stability will prevail provided

$$g \left| \frac{d\rho}{dz} \right| < \frac{8\pi p}{B^2} \left(\frac{\mathbf{k}}{k} \cdot \frac{d\mathbf{B}}{dz} \right)^2 f \left(\frac{\sigma_{\parallel}}{\sigma_{\perp}} \right), \tag{V.15}$$

where $f(\sigma_\parallel/\sigma_\perp)$ is a finite positive function of the electric conductivities and may be found by a variational method.

The accumulation of plasma in the magnetic field represents a storage of energy, not only in magnetic form, but in gravitational form as well. The storage will continue until the stability criterion (V.15) breaks down. The gravitational energy is then suddenly released, and the magnetic field will return to its previous unstressed state.

Raadu (1971) has considered the slow release of gravitational energy in a quiescent prominence, due to the slow leakage of material through the lines of force of the supporting magnetic field. The gravitational energy is released at a rate

$$G_{grav} = \rho g v_z = \mu m_H n_i g v_z, \tag{V.16}$$

where v_z is the velocity at which matter is pulled down by gravity. Raadu equated this rate to L_{rad}, the radiative loss rate from the prominence, i.e., he considered the balance

$$G_{grav} - L_{rad} = 0, \tag{V.17}$$

where L_{rad} is given by Equation (IV.13), to see what velocities v_z would be able to insure equilibrium. Rough estimates of v_z can be obtained by approximating $L_{rad} \approx n_i n_e f(T)$, and using Hirayama's (1964) expression for $f(T)$. Equation (V.17) then gives

$$v_z = \frac{n_e}{\mu m_H g} f(T) = 3 \cdot 6 \times 10^{19} \frac{n_e f(T)}{\mu}.$$

Taking $n_e = 3 \times 10^{10}$ cm^{-3}, $\mu = 1$, we find with Raadu that v_z has to be 7 m s^{-1} for a 6200 K prominence, and 1·4 km s^{-1} for an 8000 K prominence. As the temperature increases, the gravitational energy release has to increase rapidly to keep up with the radiation losses; already at $T = 10000$ K, the material will have to slip through the field lines at a speed of about 15 km s^{-1}.

This energy source for prominences has not been considered sufficiently in the literature. It shows that under certain conditions it may be of great significance, and its discussion points to the importance of dynamic models for prominences – even quiescent objects.

5.2. Dynamic Equilibrium – The Importance of Magnetic Fields

The equations governing dynamic prominence equilibria are hopelessly nonlinear, and drastic simplifications are always necessary to derive any stability criterion. We shall first present the general equations to be used and the appropriate boundary conditions. Then, in Sections 5.2.1 and 5.2.2 we consider specific prominence configuration and derive the stability criteria.

5.2.1. General formalism

The behavior of quiescent prominences in dynamic equilibrium, and the question of their stability, are studied in the hydro-magnetic approximation (i.e., neglecting high frequency effects). The equations are the ones given in Section 5.1, Equations (V.8) through (V.14), and an infinite electric conductivity is assumed.

The conditions for equilibrium, identified by subscript zero, are

$$-\nabla p_0 + \frac{1}{c}\mathbf{j}_0 \times \mathbf{B}_0 + \rho_0 \mathbf{g} = 0 \tag{V.18}$$

$$\nabla \times \mathbf{B}_0 - \frac{4\pi}{c}\mathbf{j}_0 = 0 \tag{V.19}$$

$$\nabla \cdot \mathbf{B}_0 = 0 \tag{V.20}$$

$$\mathbf{E}_0 = 0, \tag{V.21}$$

and the conditions for the magnetic field at the boundary between prominence indicated by superscript p and corona indicated by superscript c, namely:

$$\mathbf{n} \times (\mathbf{E}^p - \mathbf{E}^c) = \mathbf{n} \cdot \mathbf{v}(\mathbf{B}^p - \mathbf{B}^c) \tag{V.22}$$

$$\mathbf{n} \times (\mathbf{B}^p - \mathbf{B}^c) = \frac{4\pi}{c}\mathbf{j}^*$$

$$\mathbf{n} \cdot (\mathbf{B}^p - \mathbf{B}^c) = 0$$

$$\mathbf{n} \cdot (\mathbf{T}^p - \mathbf{T}^c) = 0,$$

become, for equilibrium

$$\mathbf{n}_0 \times (\mathbf{B}_0^p - \mathbf{B}_0^c) = \frac{4\pi}{c}\mathbf{j}^* \tag{V.23}$$

$$\mathbf{n}_0 \cdot (\mathbf{B}_0^p - \mathbf{B}_0^c) = 0 \tag{V.24}$$

$$\mathbf{n} \cdot (\mathbf{T}_0^p - \mathbf{T}_0^c) = 0, \tag{V.25}$$

where $\mathbf{n}$ is the normal to the boundary pointing out of the prominence, $\mathbf{T}$ is the stress tensor; $T_{ij} = (B^2/8\pi + p)\,\delta_{ij} - B_i B_j$ in Cartesian coordinates; and j^* pertains to the boundary surface.

The dynamic behavior of the surface is given by an equation for its unit normal vector $\mathbf{n}$ (Stix, 1962)

$$\frac{\partial \mathbf{n}}{\partial t} + (\mathbf{v} \cdot \nabla)\mathbf{n} - \mathbf{n} \times [\mathbf{n} \times (\nabla \mathbf{v}) \cdot \mathbf{n}]. \tag{V.26}$$

5.2.2. Generalized Kippenhahn-Schlüter Model

In their important paper on prominence support and stability Kippenhahn and Schlüter (1957) idealized a quiescent prominence by a plane, infinitely thin, sheet of matter standing vertically in the solar atmosphere and being supported by a magnetic field perpendicular to the filament. They studied the stability of this model against special perturbations and obtained a condition for stability, see below. More general treatments are due to Anzer (1969), Nakagawa and Malville (1969), and Nakagawa (1970).

In his work Anzer considered a generalized Kippenhahn-Schlüter model and discussed its stability against arbitrary perturbations. He used an energy principle due to Bernstein et al. (1958), modified to include configurations where the normal component of the magnetic field does not vanish at the discontinuity surfaces between plasma (prominence, indicated by superscript p) and vacuum (corona, indicated by superscript c). This modification is necessary, since it is the normal component of the magnetic field that produces the supporting Lorentz force. To show the nature of these stability analyses we shall follow Anzer's work in some detail.

With infinite electric conductivity his equations are Equations (V.8) through (V.14), except that Equation (V.14) now reads

$$\mathbf{E} + \frac{1}{c}\,\mathbf{v} \times \mathbf{B} = 0 \qquad\qquad\qquad (\text{V.14}')$$

and that the energy Equation (V.10) is reduced to $(\mathrm{d}/\mathrm{d}t)(p\rho^{-\gamma})=0$.

Anzer explored the stability of the equilibrium given by Equations (V.10) through (V.21) and subject to the boundary conditions (V.23) through (V.25). He considered small perturbations $\mathbf{r}'$ of the plasma, $(\mathbf{r}=\mathbf{r}_0+\mathbf{r}')$, and obtained the following equation in first order of $\mathbf{r}'$

$$\rho_0\,\frac{\partial^2 \mathbf{r}'}{\partial t^2} = \mathbf{F}(\mathbf{r}')\,, \qquad\qquad\qquad (\text{V.27})$$

where

$$\mathbf{F}(\mathbf{r}') = \nabla(\gamma p_0 \nabla_0 \mathbf{r}' + \mathbf{r}' \cdot \nabla p_0) + \mathbf{j}_0 \times [\nabla \times (\mathbf{r}' \times \mathbf{B}_0)] +$$
$$- \mathbf{B}_0 \times \nabla \times [\nabla \times (\mathbf{r}' \times \mathbf{B}_0)] - \nabla \cdot (\rho_0 \mathbf{r}')\,\mathbf{g}\,.$$

The disturbed fields in the corona are

$$\mathbf{E}^{c'} = -\frac{\partial \mathbf{A}}{\partial t} + \mathbf{E}_0^{c'} \qquad\qquad\qquad (\text{V.28})$$

and

$$\mathbf{B}^{c'} = \nabla \times \mathbf{A} + \mathbf{B}_0^{c'}\,, \qquad\qquad\qquad (\text{V.29})$$

where, for the vector potential, $\mathbf{A}$, we have $\nabla \times (\nabla \times \mathbf{A})=0$ and $\nabla \cdot \mathbf{A}=0$. Using Equations (V.14'), (V.22) and (V.28), we find the boundary condition for $\mathbf{A}$

$$\mathbf{n}_0 \times \mathbf{A} = \mathbf{n}_0 \times (\mathbf{r}' \times \mathbf{B}_0^{c'})\,.$$

Moving with the plasma (the prominence material) we find the changes in pressure, δp, and in magnetic field, $\delta \mathbf{B}$, to the first order in $\mathbf{r}'$

$$\delta p = -\gamma p_0 \nabla \cdot \mathbf{r}' \qquad\qquad\qquad (\text{V.30})$$

and

$$\delta B = \nabla \times (\mathbf{r}' \times \mathbf{B}_0) + (\mathbf{r}' \cdot \nabla)\mathbf{B}_0\,. \qquad\qquad\qquad (\text{V.31})$$

Anzer showed that for the disturbed interface between the prominence and the corona $\delta \mathbf{B}^{\mathrm{p}} - \delta \mathbf{B}^{\mathrm{c}} = 0$, whence Equations (V.29) and (V.31) give the boundary condition for $\mathbf{r}'$

$$\nabla \times (\mathbf{r}' \times \mathbf{B}_0) + (\mathbf{r}' \cdot \nabla)\mathbf{B}_0 = \nabla \times \mathbf{A} + (\mathbf{r}' \cdot \nabla)\mathbf{B}_0^{c'}\,. \qquad\qquad\qquad (\text{V.32})$$

The change in potential energy due to a perturbation $\mathbf{r}'$ is obtained from an integral over the whole plasma

$$\delta W_{\text{pot}} = -\frac{1}{2} \int_{\mathrm{F}} \mathbf{r}' \cdot \mathbf{F}(\mathbf{r}')\,\mathrm{d}V\,.$$

It is the behavior of δW_{pot} that determines the stability of the prominence. It is stable for displacements $\mathbf{r}'$ which gives positive values for δW_{pot}.

Anzer generalized the expression for δW_{pot} to apply to perturbations which are not restricted by constraints, like Equation (V.32). We shall denote the corresponding change in potential energy δW_1. He applied the stability analysis to the model prominence, where the magnetic field is contained in planes perpendicular to the prominence sheet (see Figure III.3). $\mathbf{B} = \{B_x(x, z), 0, B_z(x, z)\}$. The magnetic field in the corona was taken as a vacuum field, even though, as Anzer realized, a force-free field may exist.

Further, all quantities are assumed independent of the y-coordinate, and the current in the filament has only a y-component. With constant thickness X and gravitational acceleration g the condition for equilibrium, Equation (V.18), now becomes

$$\mathbf{r}' \cdot \nabla p - r_x j_y B_z - r_z \frac{\partial p}{\partial z} = 0$$

and

$$\rho g + j_y B_x + \frac{\partial p}{\partial z} = 0 .$$

By letting $X \to 0$, representing the prominence material by a sheet density $\rho(z)$ and a sheet current $j_y(z)$, and assuming, in this limit, B_x, B_z, $\partial B_x/\partial z$, $\partial B_z/\partial z$ and $\mathbf{r}'$ to be finite, we find

$$\delta W_1 = \frac{1}{2} \int \int \left\{ \frac{dB_x(x = 0)}{dz} \Delta B_z r_x'^2 - B_x(x = 0) \frac{\partial}{\partial z} \Delta B_z r_z'^2 \right\} dy \, dz , \quad \text{(V.33)}$$

where ΔB_z is the jump of B_z over the filament as $X \to 0$. The sheet current is then given by $j_y(z) = -\Delta B_z(z)$.

Conditions sufficient for stability are found from Equation (V.33) and read

$$\Delta B_z \frac{dB_x (x = 0)}{dz} > 0 \tag{V.34}$$

$$B_x (x = 0) \frac{d\Delta B_z}{dz} < 0 . \tag{V.35}$$

Anzer showed that by making these inequalities into equations, we obtain the necessary conditions for stability.

Let $\mathbf{B}^c$ be the field that existed before the formation of the filament, and indicate by $\mathbf{B}^p$ the field produced by the prominence; then $\mathbf{B} = \mathbf{B}^c + \mathbf{B}^p$ and $\mathbf{B}^c$ is current free, whence

$$\frac{\partial B_x}{\partial z} = \frac{\partial B_z^c}{\partial x} + \frac{\partial B_x^p}{\partial z}$$

and condition (V.34) becomes

$$\Delta B_z \left[\frac{\partial B_z^c}{\partial x} + \frac{\partial B_x^p}{\partial z} \right] \geq 0 . \tag{V.34'}$$

As the filament grows, the field component B_x^p increases, and so will the current in the filament. However, so long as the current is small, condition (V.34') can be approximated by $\Delta B_z(\partial B_z^c/\partial x) \geq 0$, or – since $B_x^c \Delta B_z \approx B_x \Delta B_z = \rho g > 0$ –

$$B_x^c \frac{\partial B_z^c}{\partial x} \geq 0 . \tag{V.36}$$

Condition (V.36) gives the restraint on the initial field in the corona where a prominence can form, the restraint being that the field lines must have a convex curvature pointing down where the prominence material shall accumulate. This condition was also found by Kippenhahn and Schlüter in their model.

However, as the prominence grows, the field component B_x^p may not be negligible compared to B_x^c, and the stability of the prominence against horizontal perturbations is determined by the total field **B** according to condition (V.34).

Condition (V.35) – which may be written in the form

$$\Delta B_z \frac{d}{dz} (\Delta B_z) = \frac{1}{2} \frac{d}{dz} (\Delta B_z)^2 \leq 0 ,$$

since $B_x \Delta B_z > 0$ – states that the current density may not increase with height. While measurements, see Figure II.10, seem to indicate that condition (V.34) is fulfilled in quiescent prominences, direct observational proofs of condition (V.35) are not available.

One should notice that Anzer's stability analysis holds for $X \to 0$, i.e., for an infinitely thin prominence sheet. A finite thickness introduces difficulties due to the action of the Lorentz force where the field lines have convex curvature pointing upward (in the z-direction).

5.2.3. Stability of a Horizontal Prominence-Corona Interface

In their work Nakagawa and Malville (1969) and Nakagawa (1970) considered a prominence model quite different from the Kippenhahn and Schlüter (1957) and the Anzer (1969) model. The prominence in the former case is identified with a plasma confined to the upper half plane ($z > 0$) and supported against gravity by a magnetic field **B**^{c} in vacuum. The applicability to actual prominence geometries is a little dubious, but with such a model some very interesting plasma stability conditions can be studied, and we shall discuss Nakagawa and Malville's work in some detail.

Observations of quiescent prominences on the disk often reveal a series of regularly spaced prominence 'feet' or dilations of the filament. On the limb they appear as arches and columns, giving many quiescent prominences their typical 'hedgerow' appearance. Nakagawa and Malville interpreted these regularities as the manifestation of a long-wave instability of the interface between the prominence material above and the vacuum below. The instability was assumed to be due to the shear between a magnetic field imbedded in the prominence and running along its long axis (the Y-axis), **B**^{p}, and the supporting magnetic field, **B**^{c}. The fields are initially assumed to have the form

$$\mathbf{B}_0^c = \{B_{0x}^c, B_{0y}^c, 0\}$$

$$\mathbf{B}_0^p = \{0, B_{0y}^p, 0\} ,$$

and the prominence plasma is optically thin and assumed to behave like a perfect electric conductor. With these simplifications the governing equations are again Equations (V.8) to (V.14), and the behavior of the interface can be studied by using Equation (V.26) and the boundary conditions (V.23) to (V.25).

Nakagawa and Malville noticed first that an equilibrium is possible with the following pressure distribution in the prominence

$$p_0 = \frac{1}{8\pi} [|B_0^c|^2 - |B_0^p|^2] - \rho_0 g z \tag{V.37}$$

if $\mathbf{n}_0 = \hat{z}$, where $\hat{z}$ is the unit vector in the z-direction.

Further, the authors considered small perturbations of the form $\exp i(\mathbf{k} \cdot \mathbf{r} - \omega t)$, where $\mathbf{k}$ is the wave number (k_x, k_y, k_z), leading to displacements $\delta_z = \Delta \exp i(\mathbf{k} \cdot \mathbf{r} - \omega t)$ of the prominence boundary from its equilibrium position $z=0$. To the lowest order $\Delta = (1/\omega) i v_z'(z=0)$, i.e., the vertical component of the perturbed velocity. They linearized the governing equations and derived a dispersion relation and a stability criterion for the interface.

The dispersion relation is an equation of 5th order in ω, which gives the condition for stability

$$\left(k_x^2 - \frac{\lambda}{H}\right)(k_x^2 V_A^2 + Z - \lambda g) > 0 , \tag{V.38}$$

where $\lambda = -ik_z^p$, V_A is the Alfvén velocity in the undisturbed prominence, $V_A = B_0^p / \sqrt{4\pi\rho_0}$, H is the scale height and $Z = (V_A^c)^2 (k_x \cos\theta + k_y \sin\theta)^2$.

Equation (V.38) shows that the interface is unstable for

$$k_x < \frac{\lambda}{H} , \tag{V.39}$$

i.e., for perturbations in the x-direction of wavelength longer than the scale height of the prominence material (since $\lambda \approx 1/H$). Whenever such instabilities occur, we would expect the prominence plasma to break up into more or less regular 'blobs'. Nakagawa and Malville associate the regular arch structures of many quiescent prominences (particularly of the polar crown) with these 'blobs' that occur when the inequality (V.39) is satisfied.

5.3. The Disparition Brusque as an Instability

We mentioned in Chapter II (Sections 2.2 and 2.3) that many, maybe most, low-latitude quiescent prominences at least once during their lifetime are subjected to a sudden disappearance. The whole prominence rises in the atmosphere at a steadily increasing velocity and disappears. Since the prominences often reform in the same location and basically with the same shape, it is thought that the supporting magnetic field is not destroyed, merely temporarily disturbed. This temporary disturbance seems to trigger an instability which causes the disparition brusque. Generally, the disturbances originate in flares, and aspects of the flare-disparition brusque interaction will be discussed in Chapter VI. Here we shall consider other aspects of the phenomenon that may indicate an instability in the magnetic field configuration.

5.3.1. Low-latitude filaments and polar crown filaments

While most low-latitude quiescent prominences undergo a disparition brusque, so that the phenomenon may be considered a normal event (Waldmeier, 1938), only rarely are polar crown filaments subjected to the same fate (d'Azambuja and d'Azambuja, 1948). Nevertheless, the first sudden disappearance to be photographed seems to have been a high-latitude object (Deslandres, 1897), and one of the most spectacular cases described also was a polar crown filament (Pettit, 1938).

Another important difference between high- and low-latitude filaments may be found in their response to a disparition brusque. We have mentioned that low-latitude filaments often reform. Data are scarce in this respect on polar crown filaments, but it seems that reformation generally may not take place. If this is true, it would seem natural to seek to attribute the different behavior to differences in the structure of the supporting magnetic fields for low- versus high-latitude filaments (see Hansen *et al.*, 1972).

The supporting magnetic field of a low-latitude quiescent prominence connects an area of positive magnetic polarity on one side of the prominence with an area of negative magnetic polarity on the other side. As time progresses these two areas of the bipolar configuration generally move apart, the trailing area drifting poleward. This development does not seem to impart any serious changes to the magnetic field configuration, apart from some stretching and shearing. Such modifications apparently do not inhibit the capability of the field to regain a supporting role after a disparition brusque has occurred, and the filament may reform, given the other necessary conditions for this process.

Polar crown filaments, on the other hand, are found between a poleward migrating area of a certain magnetic polarity and the 'polar cap', dominated by the opposite polarity. The supporting field bridges these areas and provides the stability for the prominence formed along the boundary between the areas. But in this case the migrating area is being pushed against the polar cap, and the supporting field – instead of having its feet in the photosphere being pushed apart leading to stretching and shearing of the field lines – will experience profound changes as the motion of the interpenetrating polarities is accompanied by partial neutralization and field annihilation.

The observation that a disparition brusque is more rare for polar crown filaments than for low-latitude quiescent prominences, may indicate that the stability of the polar crown prominence is increased over that of its low-latitude counterpart, possibly as a result of the slow neutralization process. We may speculate that if now a disparition brusque does occur, the filament is less likely to reform, since the field configuration constantly changes away from the initial conditions favorable for prominence formation.

5.3.2. Type A and type B low-latitude filaments

It is possible that a subdivision for low-latitude filaments also should be invoked. Such prominences may be found in one of the two following magnetic configurations. The filament may separate areas of opposite magnetic polarity belonging to one and the same bipolar region, i.e., the filament is found between a preceding and a following part of the region, and this may be the more common case. However, if there are two bipolar regions fairly close to each other, it is possible that the filament may be situated between the following part of one of the bipolar regions and the preceding part of the other region. The evolution of the configuration of the prominence-supporting field will be different in the two cases, and the possible ensuing sudden disappearances may be different. We shall refer to the two types of field configurations and consequently to the two types of disparition brusque as type A and type B (see Figure V.1). In type A one bipolar region is involved, for type B two such regions are required.

As type A evolves, the two parts of the plage, 1 and 2 in Figure V.1, will drift apart and, as we have mentioned, this will lead to stretching and shearing of the connecting field lines that intersect the prominence. The situation is not so clear for type B. While parts 1 and 2 will tend to separate, as will parts 3 and 4, it is not obvious how parts 2 and 3 will move in relationship to each other.

It is even possible that they may approach, in which case the prominences will be caught in a configuration not dissimilar to that of a polar crown filament.

Not much quantitative work has been done in this connection on the actual instability, or instabilities, that may cause the disparition brusque phase. Observations show that as the prominence rises, it often takes on a more or less cylindrical shape, indicating a twisted magnetic field configuration. It is known that such fields are susceptible to kink instabilities, and Hirayama (1971b) has outlined a theory along this line.

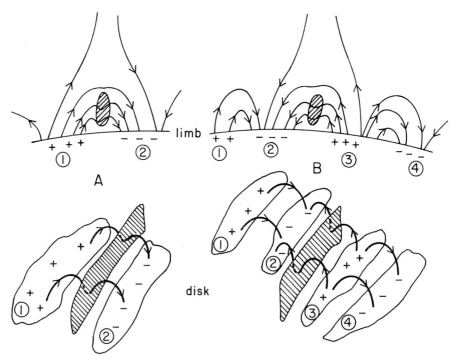

Fig. V.1. Schematic representations of two possible field configurations, type A and type B, in bipolar magnetic regions, responsible for prominence support.

CHAPTER VI

INTERACTION OF PROMINENCES WITH CENTERS OF ACTIVITY

The simplest and probably most basic form of solar activity is the magnetic plage, with a magnetic flux in excess of the flux normally channeled into the corona through the supergranulation boundaries. All prominences seem to depend for their existence or development on this excess flux, either directly or through the extra activity associated with plages. On this background the basic interaction between prominences and centers of activity is precisely this interaction with plages, which constitutes one of the prerequisites for prominences. We have often alluded to it when discussing the different types of prominences and their formation.

In this chapter we shall consider prominence perturbations and disturbances due to interactions with more severe manifestations of solar activity, notably with sunspots and flares.

6.1. The Influence of Sunspots

6.1.1. MASS BALANCE

We have mentioned several times the problems connected with the mass balance between prominences and their surroundings, and questions concerning this mass balance will continue to come up throughout the remainder of the book. Of particular interest is the effect of the mass balance on the prominence-sunspot interactions, and we shall discuss this mass budget here.

A number of active prominences are seen constantly to interact with sunspots; one observes material continuously to flow into umbral regions. Matter is guided along magnetic lines of force, seemingly falling under gravity. This motion constitutes a considerable mass loss to the prominence, and unless replenishment occurs the whole prominence will drain away often in a matter of hours, or less. However, active prominences of this sunspot type last for days, albeit with changing size and shape, indicating that some sort of mass balance is achieved. It is not obvious where the source of this matter is, but condensation out of the corona has been implied (Chapter IV.2). However, the condensation theory for this type of active prominence runs into difficulties when we consider the mass involved. We find that an unreasonably large part of the corona would be required to supply the mass for some of the more actively mass-interchanging prominences.

A way out of these difficulties may be to include some of the material shot up in the form of surges from most active sunspot groups (Chapter II.2.5), or to postulate the existence of sufficient streams of charged particles from active regions (Chapter IV.3.1). In either case the injection from below would have to be channeled into the constantly forming prominence, thence to drain into the attractive centers of sunspots.

6.1.2. Sunspot-induced prominence activations

We mentioned in Chapter V.3 that the sudden disappearance of many quiescent prominences seems to be due to an instability triggered by a perturbation from outside, from a flare or from a sunspot. Extensive studies of the nature of the flare perturbations have been undertaken, and we return to these in Section 2 below.

It is also probable that sunspots can provide the triggering agency. The work in this field is mainly due to Bruzek (1952). He found that there exists a correlation between the time, Δt, from the first appearance of a sunspot to the activation of a nearby filament, and the distance, d, of the filament from the spot. From the Δt, d correlation, Bruzek concluded that there is a perturbation propagating from sunspots with a velocity of about 1 km s^{-1}. Using observations of photospheric magnetic fields, we can link this disturbance to the development and propagation of the magnetic field around forming sunspots. The observations show that there is an expanding elliptic area around the developing sunspot group in which the chromospheric fibrils become aligned by the magnetic field. The growth of the outer boundary of this area proceeds at about 0·2 km s^{-1} (Bumba and Howard, 1965), and when the disturbance reaches a filament, the latter breaks up in a 'disparition-brusque-like' manner. This sunspot-induced disparition brusque should be clearly distinguished from the similar flare-induced phenomenon, which is triggered by a much faster disturbance, see Section 2.2. Bumba and Howard's observations show that the disturbance is to be linked with the electromagnetic field propagating away from developing sunspots. We may treat this by following Giovanelli (1947), who compared the growing magnetic field of a sunspot with the field resulting from a continuously oscillating magnetic dipole. The electromagnetic field set up around the dipole is given by (Stratton, 1941)

$$E = \mathrm{Im}\left[k_1^2 \frac{\mu}{\varepsilon'} \left(\frac{1}{r} + \frac{i}{k_1 r^2} \right) A\, e^{-i(\omega t - k_1 r)} \right], \tag{VI.1}$$

$$B = \mathrm{Im}\left[\left(\frac{1}{r^3} - \frac{i k_1}{r^2} - \frac{k_1^2}{r} \right) A\, e^{-i(\omega t - k_1 r)} \right], \tag{VI.2}$$

where ω is the frequency and A is the amplitude of the magnitude moment of the dipole, μ is the permeability of the plasma, $\varepsilon' = \varepsilon - i(4\pi\sigma/\omega)$ is the complex dielectric constant, $k_1 = (\varepsilon\mu\omega^2 - i4\pi\omega\sigma)^{1/2}$, and r is the distance from the dipole. The frequency ω is very small, that is, 10^{-6} to 10^{-5} s^{-1} for typical sunspot growth times. Giovanelli showed that the magnetic field would grow as if propagated with a velocity

$$V_{\mathrm{dist}} \approx 3(2\omega c^2/\sigma)^{1/2}. \tag{VI.3}$$

To make a rough estimate of this velocity, we take $\omega = 5 \times 10^{-6}$ s^{-1} and $\sigma = 10^8$ s^{-1}. This gives $V_{\mathrm{dist}} \approx 0·3$ km s^{-1}, not unlike the velocities inferred by Bruzek (1952) and by Bumba and Howard (1965).

The phenomenon treated here is apparently quite different from the flare-induced disturbances to be treated in Section 2.2. If the propagating magnetic field encounters a filament, the interaction between the filament magnetic field (a modified plage field) and the disturbance may possibly be quite different from the interaction between a filament and a flare-induced disturbance. Bumba and Howard's observation showed that the filament that was hit by the slow disturbance broke up over a period of two days. This may indicate that the two magnetic fields involved interacted in such a way as to profoundly alter the magnetic field configuration of the filament.

6.2. Prominence Flare Interactions

The most spectacular interaction to be discussed in the present chapter is the prominence-flare coupling. Included here is a complex set of phenomena, and several physical processes are at play. In some cases prominences are responsible for flares; these prominence-induced flarings will be treated in Section 2.1. On the other hand flares affect prominences in several ways. When a destabilization of a prominence is brought about by a flare, the result may be cataclysmic and may lead to a disparition brusque, or it may be more superficial; these prominence activations will be considered in Section 2.2. Also, some prominences are generated by flares, or appear as a result of the flare action; such loops, surges and sprays are discussed in Section 2.3.

6.2.1. PROMINENCE-INDUCED 'FLARINGS'

Bruzek (1951, 1957) has discussed in considerable detail the situation in which a disparition brusque is followed by a brightening of the chromospheric structure on either side of the site of the disappearing filament. The brightening reaches flare intensity and may last for several hours. These prominence-induced flarings were first described by Waldmeier (1938) and also mentioned by Martres (1956) and Becker (1957). According to Bruzek, the brightening of the facular structure may be described as strings of bright mottles parallel to the disappearing filament.

In cases like these, we are apparently witnessing the formation of flaring regions at the expense of energy stored in the filament (Kleczek, 1963; Sturrock and Coppi, 1965; Hyder, 1967a,b, 1968). Hyder developed a model in which modifications of the magnetic field configuration lead to a falling of the prominence material down into the chromosphere which thereby becomes heated and shows up as a brightening.

This infall-impact mechanism for solar flares has been developed further by Nakagawa and Hyder (1969), and describes how the chromosphere responds to the shock waves that develop in front of the falling material. The mechanism is capable of accounting for several of the optical, UV and X-ray emissions observed as parts of the flare phenomenon.

The infall-impact mechanism is not restricted to the disparition brusque case, but should apply to any falling material from the corona. Also, there seems to be cases of falling material whose velocity exceeds the speed of free fall, indicating electromagnetic forces in addition to gravity.

We shall now follow Nakagawa and Hyder (1969) and consider the simple case of material falling under gravity during a disparition brusque. The number density of the prominence material was taken to be $n = 6 \cdot 3 \times 10^{10}$ cm^{-3} and the temperature $T = 7 \times 10^3$ K. The supporting coronal magnetic field was assumed to be $B_c = 10$ G at a height $h_c = 30000$ km. Nakagawa and Hyder considered the material as being guided by a magnetic flux tube that passes through coronal and chromospheric layers specified by

$$n_{\text{cor}} = 10^8 \text{ cm}^{-3}, \quad T_{\text{cor}} = 2 \times 10^6 \text{ K}, \quad V_{s,\text{cor}} = 110 \text{ km s}^{-1}$$
$$n_{\text{chr}} = 10^{12} \text{ cm}^{-3}, \quad T_{\text{chr}} = 10^4 \text{ K}, \quad V_{s,\text{chr}} = 11 \text{ km s}^{-1},$$

where V_s is the sound speed. The velocity of the falling material is given by $V = \sqrt{2gh}$, where $g = 2 \cdot 7 \times 10^4$ cm s^{-2}.

After having fallen about two-thirds of the way down to the chromosphere, i.e., at $h = 10000$ km, the infall velocity will reach the sonic speed in the corona. When this occurs, a shock wave will form and will propagate ahead of the infalling material. In the beginning, because of the small Mach number of the shock, the infalling stream will continue essentially unimpeded by the shock. However, when the material reaches chromospheric levels, $h \approx 2000$ km, its velocity $V \approx 10 V_{s, \text{chr}}$, and the shock will strongly influence the stream. The reactions will lead to a decay of the shock strength and eventually to reflection of the shock wave. During this impact upon the chromosphere one would expect enhanced emission of chromospheric lines, which Nakagawa and Hyder identify with the optical flare. With the free fall considered in this model, the impact upon the chromosphere should occur about 7 min after the start of the infall, i.e., after the beginning of the disparition brusque phase, in reasonable agreement with some observations.

6.2.2. DESTABILIZATION OF PROMINENCES

(a) *Introductory Remarks*

Many flares generate during their most active phase (the flash phase) a wave disturbance that propagates at great speed (around 1000 km s^{-1}) in the corona and that may destabilize quiescent prominences in its way. In severe cases the prominence-supporting magnetic field becomes affected to such a degree that the filament undergoes a disparition brusque. In other instances the filament may be set in damped oscillations, after which it again regains its stability. During the oscillations, which apparently may be either mainly horizontal or mainly vertical, the filament may give a spectrum whose lines are Doppler shifted relative to an observer so that the image of the filament alternately is shifted out of and into the passband of narrow-band Hα filters used for flare patrols. This effect gives a 'winking' impression of the image, and the phenomenon then is referred to as winking filaments.

These winking filaments have been known for many years. The phenomenon was observed in spectrohelioscopes by Greaves, Newton and Jackson (Dyson, 1930) and studied further by Newton (1935). The observed winking filament generally shows an initial receding motion followed by from one to four damped oscillations. The frequency of oscillations varies from one filament to another, and does not appear to relate to the importance of the flare. Ramsey and Smith (1966) observed a filament which was disturbed four times in a three-day period, and each time it oscillated with essentially the same frequency. There is further a preferential direction, or rather a cone, in which the disturbance propagates. Filaments lying inside this cone as seen from the flare may be activated, those outside are not.

The work on these activations was continued by Dodson-Prince (1949), Bruzek (1951, 1958), and Becker (1958) and especially by Moreton (1960, 1965), whose refined photographic technique permits the actual observation of the propagating disturbance as it travels from the flare to the filament with a velocity in the range 500 to 1500 km s^{-1}. For other significant contributions to the observations of these activations, see papers by Moreton and Ramsey (1960), Dodson-Prince and Hedeman (1964), Smith and

Angle (1969). Theoretical investigations have been made by Athay and Moreton (1961), Anderson (1966), Hyder (1966), Meyer (1968), Uchida (1968, 1970) and Kleczek and Kuperus (1969).

The nature of these flare-generated disturbances will be discussed in Section 2.2.d below. In Sections 2.2.b and 2.2.c we shall consider two different ways of analyzing the prominence oscillations, as vertical or as horizontal oscillations. In both cases the motions may be treated with the equation for a damped harmonic oscillator. The observations of Ramsey and Smith (1966) have been used to infer the frequency of the oscillation, v_{osc} (or the period P) and the damping time τ (or the decay constant γ). Ramsey and Smith used narrow-band observations of the filament in the center of the Hα line, and at $+0.5$ A and -0.5 Å from the line center, and deduced the above-mentioned parameters from the relative visibility of the filament at these three wavelengths.

(b) Model for Vertical Oscillations

Malville (1961) suggested that the oscillatory motions could probably be understood in terms of the magnetic fields in the filaments, and Hyder (1966) developed a model where an electromagnetic wave disturbance causes the filament to oscillate under the influence of its own magnetic field (see also Anderson, 1966). We shall follow Hyder's arguments in some detail, taking the frequency of oscillation to be $v_{osc} = 10^{-3}$ s^{-1} and the decay constant $\gamma_z \approx 10^{-3}$ s^{-1}. The mass of the filament is supposed to be suspended in the magnetic field supporting the quiescent prominence, in a manner similar to the model of Kippenhahn and Schlüter (Chapter III.1.3.b). The vertical component of the field is B_z, the scale height of the filament is H and a small downward displacement Δz of the filament is supposed to lead to a linear increase in B_z,

$$\Delta B_z = -B_z \Delta z/H . \tag{VI.4}$$

Under these conditions, one can analyze the filament oscillations in terms of a damped harmonic oscillator, and the frequency is determined from the equation of motion

$$\frac{d^2 z}{dt^2} + \frac{\mu}{M}\frac{dz}{dt} + \frac{K_z}{M} z = 0 \tag{VI.5}$$

and from

$$v_{osc} = \sqrt{v_0^2 - \mu^2/16\pi^2 M^2}, \tag{VI.6}$$

where v_0 is the frequency of an undamped harmonic oscillator, $v_0 = (1/2\pi)\sqrt{K_z/M}$; M is the mass of the filament, K_z the restoring force, and μ the coefficient of friction in the corona in which the filament moves. The ratio K_z/M can be expressed in terms of the magnetic field in the prominence (through the magnetic tension, $B^2/4\pi H$) and the physical parameters describing it (density ρ, volume V, scale height H):

$$\frac{K_z}{M} = 2 \frac{d}{dz}\left(\frac{B_z^2}{4}\right)\frac{V}{H}\frac{1}{\rho V} . \tag{VI.7}$$

Combining Equations (VI.4) and (VI.7), we find

$$\frac{K_z}{M} = \left(\frac{B_z}{H}\right)^2 \frac{1}{\pi\rho} . \tag{VI.8}$$

This equation may be combined with the following expression for K_z/M in terms of the frequency of oscillation and the decay constant, that is,

$$K_z/M = 4\pi^2 v_{osc}^2 + \gamma_z^2$$

to obtain an expression for the vertical field

$$(B_z/H)^2 = \pi\rho(4\pi^2 v_{osc}^2 + \gamma_z^2) \, .$$

The frequency v_{osc} as given by Equation (VI.6) depends on the friction experienced by the oscillating filament. The coefficient of friction may be defined in terms of the coefficient of viscosity in the corona $\eta = 2M \, d\gamma/Ac$, where A is the area of the vertical surface of the filament and d the effective distance, perpendicular to A, over which shears exist in the coronal plasma as a result of the vertical oscillations. Hyder inserted the following values for the parameters: $M = 10^{15}$ g, $H = 3 \times 10^9$ cm, $d = 10^9$ cm, $A = 10^{20}$ cm^2, which leads to

$$B_z \approx 10 \, G \, , \quad \eta \approx 10^{-9} \text{ poise} \, .$$

Linhart (1960) has given a relation between the strength of a coronal magnetic field, B_c, and the coefficient of viscosity in the corona

$$\eta = 1.6 \times 10^{-26} \, n_e^2/B_c^2 T_c^{1/2} \, .$$

Using the value for η determined by Hyder and inserting for $n_e = 10^9$ cm^{-3} and for $T_e = 10^6$ K, we find $B_c \approx 0.13$ G. The quoted values of B_z and B_c indicate that an enhancement of the field strength between one and two orders of magnitude is involved in the formation of a quiescent prominence.

Hyder implies that the disparition brusque and the oscillating filament are similar phenomena. The property that distinguishes between them may be the direction of the initial displacement Δz of the filament. If the displacement takes place along the positive z-axis, the filament may completely disappear, but if the displacement is downward (along the negative z-axis), the disturbed filament may undergo oscillations. As the filament is pushed down initially, the density increases rapidly and the motion becomes highly damped.

(c) Model for Horizontal Oscillations

According to Kleczek and Kuperus (1969) observations indicate that most prominence oscillations are horizontal, rather than vertical, i.e., that the prominences – regarded as vertical sheets – execute horizontal, periodic motions from and towards the disturbing flare. In this picture we can think of the sheet-like prominence being hit broadside by the flare disturbance and displaced horizontally from its equilibrium position. Kleczek and Kuperus approximated the supporting magnetic field by an effective field along the long axis of the prominence (the y-axis) and anchored in fixed positions. The horizontal restoring force, K_x, then is given by the magnetic tension $-(1/4\pi)(\mathbf{B} \cdot \nabla)\mathbf{B}$. If $2L$ denotes the length of the prominence and x the displacement at the middle of it, the restoring force is to first order proportional to x and given by $(B_y^2/4\pi L^2)x$. The effect of this force is to induce damped oscillations of the prominence sheet according to the equation

$$\frac{d^2x}{dt^2} + \frac{R}{M}\frac{dx}{dt} + \frac{K_x}{M}x = 0 , \tag{VI.9}$$

where R is the damping constant. As alluded to in the introductory remarks Equation (VI.9) is mathematically the same equation that Hyder used (Equation (VI.5)) but in Hyder's case of vertical oscillations the damping is due to magnetic viscosity. For a sheetlike prominence the effect of viscosity is negligible under horizontal oscillations.

However, as pointed out by Kleczek and Kuperus, the coronal plasma will alternately be compressed and rarified during the horizontal oscillations, and compression waves will be generated. The authors analyzed this situation in terms of the theory for radiation of acoustic waves from a circular piston (Lindsay, 1960). The emission of waves of the type $(1/r)\,e^{i(\omega t - kr)}$ from a circular piston of area πa^2 will affect the motion of the piston, and the effect can be described as a radiation reaction force R. If we neglect the acoustic reactance part of this force – which basically results in an increase in the effective mass of the prominence, and therefore in a, relatively small, increase in the oscillation period – R may be written

$$R = \pi a^2 \rho_c c \frac{dx}{dt}\left[1 - \frac{J_1(2ka)}{ka}\right], \tag{VI.10}$$

where ρ_c is the coronal density and $J_1(x)$ the Bessel function of first order. Equations (VI.9) and (VI.10) combine to give the following equation of motion for the damped oscillating prominence

$$\frac{d^2x}{dt^2} + \frac{\rho_c c \pi a^2}{M}\left[1 - \frac{J_1(2ka)}{ka}\right]\frac{dx}{dt} + \frac{a^2 B_y^2 X}{4ML^2}x = 0 , \tag{VI.11}$$

where $\pi a^2 = 2ZL$ is the surface area of the prominence sheet of height Z and thickness X. Equation (VI.11) gives the oscillation period $(P = 2\pi/\omega = 2\sqrt{M/K_x})$

$$P = \frac{4\pi L}{M}\sqrt{\pi \rho_p} , \tag{VI.12}$$

where ρ_p is the density of the prominence plasma $(= M/2LZX)$. Kleczek and Kuperus took $2L = 10^{10}$ km, $\rho_p = 10^{-14}$ g cm^{-3} and $B = 9$ G to find a period of about 20 min. The decay constant for the oscillations, $\gamma_x = R/M$, is given by

$$\gamma_x = \frac{\rho_c c A}{M}\left[1 - \frac{J_1(2ka)}{ka}\right]. \tag{VI.13}$$

For two cases studied by Kleczek and Kuperus, that gave $P = 20$ min, Equation (VI.13) indicates a decay constant of about 4×10^{-4} s^{-1}.

(d) The Flare Disturbance

The nature of the perturbations that emanate from flares and cause the prominence disturbances has been the subject of numerous investigations. We have mentioned that the perturbations propagate with velocities of the order of 1000 km s^{-1}; however, the spread is large, between less than 400 km s^{-1} to more than 1800 km s^{-1}. The propagation takes place in the chromosphere or the corona at speeds much greater than the velocity of sound, V_s. For comparison, $V_s = 23$ km s^{-1} for a temperature of

$T = 2 \times 10^4$ K, and in the corona V_s is of the order of 200 km s^{-1}. However, the Alfvén velocity in the corona probably ranges from 1000 to 3000 km s^{-1}. It is necessary to keep these values in mind when we now discuss the nature of the disturbances.

Explanations in terms of corpuscular ejections from flares or shock waves of high Mach number, propagating through the chromosphere, fail to satisfy the observations (Uchida, 1968). Also, we do not expect the gravitational field to affect the disturbances significantly, as seen from the following discussion.

The linearized theory for adiabatic oscillations in an isothermal plane-parallel atmosphere due to pressure gradients, gravity, and inertia leads to a dispersion relation

$$\omega^4 - \omega^2 V_s^2(k_x^2 + k_z^2) + (\gamma - 1)g^2 k_x^2 + ig\omega k_z = 0, \tag{VI.14}$$

where we have sought solutions of the form $\exp i(\omega t - k_x x - k_z z)$ for the perturbation velocity, pressure and density.

If we neglect gravity, Equation (VI.14) reduces to $\omega^2 = V_s^2(k_x^2 + k^2)$, which is the dispersion relation for ordinary sound waves when k_x and k_z are real. When gravity is present, there is no solution of Equation (VI.14) for which both k_x and k_z are real. There is one solution for which k_x is real, and k_z has the form $k_z = k_z$ (real) $+ i\gamma g/2V_s^2$. This solution permits the phase to change in the vertical direction; hence, propagation is possible. Thus these waves include sound waves in a generalized form which show the effect of gravity, and, as we shall presently see, they also include the low frequency gravity waves.

Introducing the expression for the complex wave number k_z in Equation (VI.14), we find the dispersion relation

$$\omega^4 - \omega^2 V_s^2(k_x^2 + k_z^2) + (\gamma - 1)g^2 k_x^2 - \gamma^2 g^2 \frac{\omega^2}{4V_s^2} = 0.$$

This equation shows that for each value of the wave number, there are two positive values of ω, that is, two waves modes. For one mode, we have

$$\omega > \omega_a \equiv \gamma g/2V_s, \tag{VI.15}$$

and for the other, we have

$$\omega < \omega_g \equiv g\sqrt{\gamma - 1}/V_s. \tag{VI.16}$$

The high frequency mode whose frequency is greater than the critical frequency ω_a is the acoustic wave mode, whereas the low frequency mode, found only for frequencies smaller than the critical frequency ω_g, is the (internal) gravity wave mode. The acoustic waves include ordinary sound waves as a limiting case. Furthermore, we see from Equations (VI.15) and (VI.16) that the domains of the acoustic waves and of the gravity waves are completely separated, since there is no real value of γ smaller than 2 for which ω_a and ω_g coincide.

When we allow temperature variations in the atmosphere, the wave modes just described still retain their identity; that is, for each value of the wave number we have a high frequency acoustic wave and a low frequency gravity wave. But whereas the waves propagate in straight lines in the isothermal case, we encounter curved wave paths as the result of a temperature gradient.

Observations show that the dominant wave frequency of the disturbance is of the order of 0·3 rad s⁻¹, while at chromospheric temperatures $\omega_a \approx 0\cdot02$ rad s⁻¹. Consequently, we should clearly be in the acoustic mode.

The most likely explanation for the nature of the disturbances is that we are faced with weak hydromagnetic shocks generated during the flash phase of the flare (Anderson, 1966; Uchida, 1968, 1970). Anderson considered a disturbance-pulse with the form of an N-wave, exhibiting the rapid rise in pressure – the head shock – followed by a rarefaction wave in which the pressure drops linearly below the ambient value an (absolute) amount equal to the initial discontinuity, and then followed by a sudden rise to the ambient level – the tail shock. This N-wave is now guided along the magnetic lines of force up into the corona, whence it eventually is reflected and refracted back into the chromosphere. Anderson assumed the propagation to take place in a magnetic field approximated by that of a magnetic dipole situated below the photosphere. By using a ray-tracing technique (Haselgrove, 1954, 1957; Budden, 1961), he then determined the path of the weak shock and the time of arrival at different distances in the chromosphere away from a flare source which was placed above the magnetic dipole. For instance, the disturbance will arrive in the chromosphere at a distance of 140 000 km from the source after about 150 s. From the observed activation of a prominence situated at this distance, we deduce a perturbation velocity of slightly less than 1000 km s⁻¹.

The assumed model for the magnetic field in the corona is idealized and will not describe actual field configurations too well. Also, the choice of the particular wave mode and the shape of the shock is somewhat arbitrary, but the idea of using slightly nonlinear hydromagnetic waves to describe the disturbance is interesting, and further studies along these lines may explain other aspects of the disturbance (directivity of the disturbance, interaction with spicules, etc.).

Uchida (1968, 1970) used a canonical-type formulation for the propagation of hydromagnetic disturbances, in a non-uniform, anisotropic corona in the WKB approximation, to calculate the loci of re-entry into the chromosphere of the wave-packets. He showed that in this approximation the propagation of the slow mode and the Alfvén mode wave-packets is anisotropic, taking place along the magnetic field lines, whereas the propagation of the fast mode wave-packet is isotropic. Uchida chose the fast mode, arguing that it seems very unlikely the magnetic field lines – necessary to guide the slow and Alfvén modes – will have the right configuration to ensure propagation from the flare in question to all the observed distant locations in the chromosphere. Furthermore, the slow mode is just too slow. For the fast mode he computed the motion of the wavefront in the corona by assuming different coronal models (magnetic field and density distributions), see also Uchida *et al.* (1972), and showed that solutions could be found for which the wave packets move out from the flare, at observed velocities, due to the difference in flight time for different ray altitudes.

6.2.3. PROMINENCE GENERATION

Certain types of prominences seem to be generated directly by the flare, i.e., loops and sprays, and these two names should be used only in conjunction with flare-associated prominences. On the other hand, as far as we know, surges may or may not be flare-connected. Most of the phenomena mentioned in this section have been considered earlier in the book. Consequently, we shall here only mention some aspects of the prominences pertaining specifically to their interaction with flares.

(a) *Ejections*

Bruzek (1969b) distinguishes several types of ejections from flares (see also Chapter II.2). These ejections are flare material shot out from the main body of the flare – they may even constitute the main part of the flare. That we are dealing with flare material may be inferred from the optical spectrum of surges, which resembles the flare spectrum rather than the spectrum of quiescent prominences (including eruptive prominences).

We distinguish between sprays (which include Bruzek's fast ejections and sprays), and the slower and less spectacular flare surges. This may be only a practical distinction; we do not know whether there is any physical difference in the generation mechanisms involved – even though there is considerable difference in the energy requirements in the two cases. In any event we require a mechanism that can utilize at least part of the available energy in an acceleration process for part of the flare plasma. For a large surge the mechanical energy needed is of the order of 10^{31} erg (Bruzek, 1969b).

Since surges have a strong recurrence tendency, it is likely that the acceleration mechanism does not demand destruction of the energy source. If a magnetic field is involved, we are witnessing a redistribution to a lower energy state, rather than an annihilation of the local field.

(b) *Loops*

While all loops are flare associated, there are flares that do not generate loop systems. However, here we should be careful and define more precisely what we mean by a flare. Traditionally, a flare is defined as an excess brightening observed in $H\alpha$, e.g., a low-energy plasma. But this is only part of the story in many cases. The observation of EUV emission, X-rays, radio bursts and fast particles from many flares testifies to another physical regime: a high-energy plasma. It is possible that it is this part of the flare that is associated with the loop generation; in other words only flares that develop a high-energy plasma are accompanied by loops. A high correlation exists between occurrence of loop systems and emission of fast protons by flares (Bruzek, 1964a, b; Jefferies and Orrall, 1964).

Alternatively, it may be that the loop system should not be considered secondary to the flare itself, but that there is a common cause responsible for both. The optical spectra of loops and limb flares are nearly identical, a fact again testifying to the close relationship between the two phenomena. According to this alternative loops and flares are but two manifestations of a single, albeit complicated, activity; two related, if slightly different, responses to an instability of an active region in the corona. The instability may sometimes trigger both a flare and a loop system, sometimes only the flare.

6.3. Interaction of Prominences with Each Other

Most prominences may be considered as single objects and may be described and understood adequately as such. However, there are others whose interaction with another prominence constitutes a most interesting phase of their life, and their

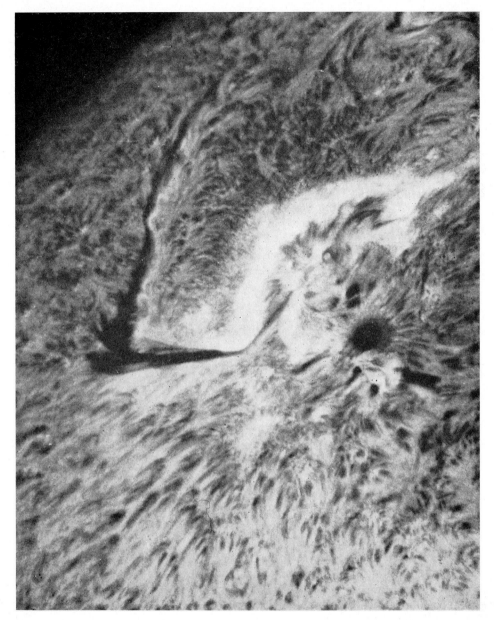

Fig. VI.1. Filament with an active part in an active region and a quiescent part extending out through a quieter part of the plage (courtesy S. F. Martin, Lockheed Solar Observatory).

description would not be complete without incorporating this interaction. Many examples could be cited, we shall treat only a few types to illustrate this behavior.

6.3.1. FIBRIL-FILAMENT INTERACTIONS

Observations with high spatial resolution indicate that the formation of both quiescent and active filaments is intimately linked to the configuration of pre-existing chromospheric fibrils, the smallest structures we recognize as prominences. The first step in the formation of an active region filament is the alignment of the fibrils (S. Smith, 1968). It seems that this alignment provides a necessary condition for the filament to form, either by making possible condensation of material onto this configuration, or by providing a channel for material to flow into it. We have mentioned earlier (Chapter II.2) that material always is seen to flow along the long axis of the filament.

If the filament grows out of the active region it may form a quiescent prominence. The fine structure in such objects is more vertical than horizontal, and motions are observed to follow the same pattern. This behavior probably is due to the fact that the filament now finds itself in magnetic fields significantly weaker than the field in the active part of the plage. It is not uncommon to see a filament with an active end in an active region and a quiescent part extending out through a quieter part of the plage (see Figure VI.1).

6.3.2. COLLIDING PROMINENCES

We have mentioned that material often is seen to stream into or out of a prominence, thereby making it interact with a sunspot, a coronal cloud, etc. At times a much more violent interaction takes place, and we may witness the collision of two prominences. In such cases it is possible to explosively release energy and profoundly change the prominences in question.

Kleczek (1969) observed a surge-like explosion being captured by overlying loops. As a result the loops were stretched and distorted, and from the distortion he estimated the magnetic field in the loop that was capable of stopping the surge motion.

It is interesting to note that Roberts and Billings (1955a, b) considered the kinetic energy associated with such surge motions as a possible source of flares. They suggested that the surge may be trapped by the loop-like magnetic field configuration, a sudden compression of the surge plasma then will occur, and an explosive conversion of the kinetic energy to heat might result.

In Kleczek's case the kinetic energy of the surge, $\varepsilon_k = \frac{1}{2}MV^2$, where M is the mass of the surge and V its gross velocity, is equal to the increase in magnetic energy due to the stretching of the field lines, $\Delta\varepsilon_B = (B^2/8\pi)\,\Delta lA$, where Δl is the lengthening of the flux tube of cross-sectional area A. Equating ε_k and $\Delta\varepsilon_B$, we find

$$B = \sqrt{\frac{4\pi MV^2}{\Delta lA}}\,. \tag{VI.17}$$

Kleczek inserted reasonable values in Equation (VI.17) for the size and motion of surges and tried to measure Δl. As a result, he deduced magnetic field strengths in the range 50 to 150 G, in general agreement with observed fields in loop systems.

CHAPTER VII

PROMINENCES AS PART OF THE CORONA

Prominences are found in the corona, but generally are considered as small perturbations and therefore neglected in most considerations of coronal physics. However, there may be several aspects of the corona that are poorly treated if the influence of prominences is ignored completely. Already Lockyer (1903, 1922, 1931) showed that there is a close association between the distribution of prominences on the disk and the form of the corona. In this chapter we shall look at several prominence-corona associations, mainly on a short time scale. The list is hardly exhaustive, but some important questions in coronal and interplanetary physics will be brought into focus.

7.1. Quiescent Prominences

The role of quiescent prominences in coronal physics is illustrated by a discussion of the mass balance of the corona and by a consideration of the shape and locations of coronal 'cavities' and streamers.

7.1.1. Mass balance

With Athay (1972) we may compare the mass of quiescent prominences with the mass of the whole corona. We obtain a measure of the latter by the expression

$$M_{cor} \approx n_p m_p A H, \qquad (VII.1)$$

where the proton density $n_p \approx 3 \times 10^8$ cm^{-3}, the area $A = 4\pi R_\odot^2 = 6 \times 10^{22}$ cm^2 and the scale-height $H \approx 10^{10}$ cm. Inserting these values in Equation (VII.1), we find $M_{cor} = 3 \times 10^{17}$ g.

Similarly, the mass of a large quiescent prominence is approximately

$$M_{prom} \approx n_H m_H V, \qquad (VII.2)$$

where the hydrogen density n_H is determined by the condition for pressure equilibrium with the corona, i.e., $k n_H 6000 = k(n_p + n_e) 2 \times 10^6$, or $n_H \approx 3 \times 10^{11}$ cm^{-3}, and V is the volume, about 10^{29} cm^3. With these values, Equation (VII.2) yields $M_{prom} \approx 5 \times 10^{16}$ g $\approx 0.2\ M_{cor}$.

Such order of magnitude estimates indicate that half a dozen large quiescent prominences are as massive as the whole corona. Consequently, a single quiescent prominence constitutes a significant part of the mass stored in the corona at any one time. This fact has a bearing on the theory of formation of filaments. If they condense out of the corona (Chapter IV.1), material must continuously be supplied to the adjacent coronal regions from denser regions below (maybe in the form of spicules, surges, etc.).

10*

7.1.2. Coronal cavities

When quiescent prominences are observed end-on (parallel to their long axis) above the solar limb at eclipses, one generally finds that the inner corona shows a region of reduced brightness surrounding the filament. This dark area is referred to as a cavity. Such observations date back to the eclipse of January 22, 1898 (Wesley, 1927), and the phenomenon has been studied many times since then (Waldmeier, 1941, 1970; van de Hulst, 1953; von Klüber, 1961; Williamson *et al.*, 1961; Kleczek and Hansen, 1962; Leroy and Servajean, 1966; Kawaguchi, 1967; Saito and Hyder, 1968).

The effect is observed also in the monochromatic emissions of coronal lines. Waldmeier (1946, 1952, 1970) used this information to show that the reduced brightness cannot be explained in terms of decreased temperature (both the $Fe x$ 6374 Å and the $Fe xiv$ 5303 Å lines are weakened), and the cavity must owe its existence to a reduced electron density (see also Waldmeier, 1941).

Closely related to the cavities are the so-called coronal arches. These form a system of nearly concentric arches around the prominence. The arches are often elliptical in shape and form the lower region of helmet-shaped streamers (Balanovsky and Perepelkin, 1928; Wallenquist, 1957; Leroy and Servajean, 1966; Kawaguchi, 1967; Saito and Hyder, 1968). While the dimensions of the cavity are greater than the dimensions of the prominence only by about a factor of 2, the arches can be seen out to about $0.5R_0$ from the limb. The width of an arch increases with height, but 40000 km may be an average value. The brightness of the cavity is about 20% of the brightness of the lowest arch. The connection – if any – between the existence of the cavity and surrounding arches on one hand and the quiescent prominence on the other, has not been studied. There are implications for the theory of condensation of prominences that should be investigated.

7.1.3. Model of the prominence-streamer association

According to Pneuman (1972) both the prominence and the surrounding cavity may be considered consequences of the adjustment required in the density structure of the helmet to satisfy the energy balance as modified by the magnetic field. Areas of the corona where the magnetic field is open (connecting with the outer corona) are subjected to heat conduction losses and possess a temperature decreasing with height. On the other hand, one expects regions where the field lines are closed (return to the surface in the lower corona) to be nearly isothermal, due to the inhibition of outward heat conduction. Consequently, the scale height falls off more rapidly in open regions than in the closed field regions, and there results a relative increase in density with height in the closed regions (Pneuman and Kopp, 1970). It is in such areas that we find the base of helmet streamers. We shall follow Pneuman's argument and see how such regions may adjust to the energy transport.

Consider a flux tube extending from a photospheric region, A, into the corona, reaching maximum height at B, and bending down into the photosphere again at C. Suppose that the temperature, which first increases with height along the field lines to some maximum value, T_{max}, and then decreases further out, reaches T_{max} somewhere between the levels A (or C) and B. We take this level of the temperature maximum as the reference level, indicated by a suffix zero. Pneuman considered the case where the steady state in the flux tube is brought about by having the total energy input (in the form of wave energy) at the base of the flux tube being radiated away within its volume, or, from Equation (IV.20)

$$G_{\text{mech}} - L_{\text{rad}} = 0 . \tag{VII.3}$$

Thermal conduction is not included in Equation (VII.3). The conduction is prohibited across the field lines of the tube, and since the base of the flux tube is at the temperature maximum, the temperature gradient vanishes there. Hence, there is no conductive flux across dA_0 either. As a result, conduction can serve only to redistribute energy within the volume of the tube, and the temperature will be very nearly uniform.

For G_{mech}, instead of using expression (IV.13), Pneuman writes the term in the form

$$G_{\text{mech}} = 2F_{\text{mech}} \, dA_0 , \tag{VII.4}$$

where F_{mech} is the mechanical energy flux entering the tube at the coronal reference level, and dA_0 is the cross-sectional area of the tube. The radiation loss is given by Equation (IV.14), or

$$L_{\text{rad}} = \text{const } n_e^2 T^\alpha ,$$

where α has the value -1 in the temperature range above 10^6 K (see Table IV.1).

The above mentioned assumptions and Equations (IV.14), (VII.3) and (VII.4) combine to give

$$\int \int_V \int \text{const } n_e^2 T^\alpha \, dV = 2F_{\text{mech}} \, dA_0 , \tag{VII.5}$$

where we integrate over the volume within the flux tube.

Pneuman considered the gas in the flux tube at rest and used the equation of hydrostatic equilibrium along the field lines to relate the density at height h to the parameters at the reference level, or

$$n_e(h) = n_{e,0} \, e^{-h/H} = n_{e,0} \, e^{-(r_0/H)(1-r_0/r)} , \tag{VII.6}$$

where H is the scale height, and r the radial distance. Since T is uniform in the flux tube, Equations (VII.5) and (VII.6) give a relation between the density at the base level of the tube and the temperature and mechanical energy flux for a given magnetic field configuration

$$\int \int_V \int e^{(2r_0/H)(1-r_0/r)} \, dV = \frac{2F_{\text{mech}} \, dA_0}{\alpha n_{e,0}^2 T^\alpha} . \tag{VII.7}$$

The simple field configuration due to a dipole in the Sun's center is sufficient to bring out Pneuman's main idea. The volume element is $dV = dA(s) \, ds = (B_0/B) \, dA_0 \, ds$, where ds is an element of arc length along the field line, and where we have used the condition for magnetic flux conservation along the flux tube $(B(s) \, dA(s) = B_0 \, dA_0)$. For the dipole field, the total field strength at the reference level is $B_0 = B_p \sqrt{(\cos^2 \theta_0 + \frac{1}{4} \sin^2 \theta_0)}$, where θ_0 is the polar distance and B_p is the field strength at the pole $(\theta_0 = 0)$. If B_r is the radial component of the field, then $ds/B = dr/B_r$. Combining the above expressions, we can write – after some rearrangements – Equation (VII.7) in the form

$$1 - e^{-(2r_0/H)(1/\sin^2\theta_0 - 1)} = \frac{2F_{\text{mech}}}{Hn_{e,0}^2 T^\alpha} \sqrt{\frac{1 - \sin^2\theta_0}{1 - \frac{3}{4}\sin^2\theta_0}}.$$ (VII.8)

To specify $n_{e,0}$ and T separately, Pneuman considered the balance of forces across the magnetic field lines, and required that $p + B^2/8\pi = \text{const}$. For the dipole field distribution this assumption can be written

$$n_{e,0}kT = \text{const} + \frac{B_p^2 - B^2}{8\pi} = \text{const} + \frac{3B_p^2}{32\pi}\sin^2\theta_0.$$ (VII.9)

He could then solve Equations (VII.8) and (VII.9) simultaneously to obtain $n_{e,0}$, and find the temperature distribution from Equation (VII.9). Finally, the general density distribution $n_e(h)$ can be found from Equation (VII.6).

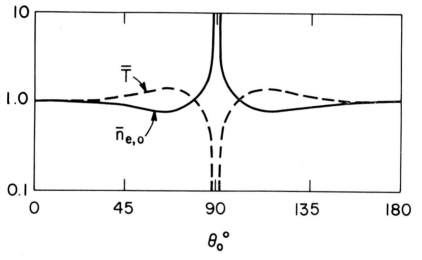

Fig. VII.1. Example of density (solid curve) and temperature (dashed curve) profiles across the base of a coronal streamer, as calculated by Pneuman (1972) from Equations (VII.8) and (VII.9). Density and temperature are normalized to the values at the pole ($\theta_0 = 0$), i.e., $\bar{n}_{e,0} = n_{e,0}/n_p$, $\bar{T} = T/T_p$.

Typical solutions for $n_{e,0}$ (solid curve) and T (dashed curve), both normalized to their value at $\theta = 0$, are shown in Figure VII.1 across the base of a streamer situated at the equator ($\theta = 90°$). The density is fairly uniform over most of the coronal structure, but decreases as we approach its center. Finally, very near the center the density rises rapidly. Pneuman identifies the region of lower density (and higher temperature) with a coronal cavity, and the region of high density (and very low temperature) at the center with a quiescent prominence.

Pneuman's model places the prominence above the temperature maximum of the corona. The height of this maximum is not known, and it would be of great interest to determine it, and compare with observed lower boundaries of quiescent prominences.

7.2. Active and Activated Prominences

The overall influence of the many types of active prominences on the corona is difficult to assess. The contribution of individual surges to the mass of the corona for example may be subtle, but it is not doubtful that at times of great activity considerable mass exchange takes place over active regions. Observations above the limb often reveal pronounced and prolonged coronal rain which provides an important drain of coronal matter. Similarly, disk observations show that as an active region develops, there is a nearly continuous streaming of matter down into the region (Ramsey *et al.*, 1968).

Following many important flares – or appearing concurrently with such flares – we may observe two very different types of highly active prominences, viz., loops and sprays. Both types strongly influence the above-lying corona, but in distinctly different ways, as discussed below.

Also quiescent prominences may – following periods of activation (Chapter V.3) – profoundly influence local areas of the corona. This influence is due to the fact that the activated prominence – in its disparition brusque phase – ascends in the corona – as do sprays. We shall refer to the complex transitory responses to any kind of ascending prominence as coronal transients, and we treat them in Section 2.2 below, after having discussed the more stationary – albeit highly energetic – association between loops and the corona in Section 2.1.

We note that the term coronal transient is not well defined. It has been used in a more restricted sense to describe temporary changes in the intensity distribution of the white-light corona. However, these temporary changes are not isolated phenomena, but seem to be intimately related to ascending material and to generation of radio bursts due to the passage through the corona of waves and particles.

7.2.1. LOOPS AND CORONAL CONDENSATIONS

One of the strongest prominence-corona associations is provided by loop prominence systems and the accompanying coronal condensations. Newkirk (1957) showed that the gross motion of the coronal plasma, V_{cor}, near the top of the loops is similar to the motion of the loops, V_{prom}. The velocities have the same sign, but in the mean $V_{cor} \approx \frac{1}{3} V_{prom}$.

We have seen (Chapter III.3.3.b) that the shape of loop prominences seen in Hα corresponds well with the configuration of magnetic field lines as computed from potential field theory, using photospheric magnetic-field data (Harvey, 1969). Rust and Roy (1971) have shown that also the coronal loop structures, seen in the Fe xiv, 5303 Å line above active region – and being nearly identical to the Hα loops photographed at the same time – closely follow the shape of computed potential magnetic field lines.

Spectra indicate that the prominence and coronal loops contain densities of $10^9 - 10^{11}$ cm^{-3}, and the temperature ranges from 4×10^4 K to 3×10^7 K (Newkirk, 1971). Furthermore, the tops of the prominence loops, which are closer to the coronal condensation, are more 'active' than the lower parts of the prominence (Chapter II.1.3.c).

We conclude from this that the active corona above and around loop prominence systems is intimately coupled to the loops, and this coupling extends to the flare activity that always precedes the formation of the loop systems (see also Bruzek and DeMastus, 1970). The physical mechanism involved is not known, but its nature poses one of the outstanding questions in the theory of active prominences.

7.2.2. ASCENDING PROMINENCES AND CORONAL TRANSIENTS

The other highly active type of prominence which we shall consider in the context of the interaction with the corona, is the spray. Relative to the organized, well predictable development of loops, sprays form their counterpart of chaotic behavior. Several examples of spray prominences, which all are closely flare-related, have been discussed in the literature (Chapter II.2). However, due to their high velocity many sprays must have gone unnoticed. Also, we do not know much about their interaction with the coronal plasma. To remedy this, Øhman (1971a,b) instigated an ambitious spray patrol program that is providing much-needed valuable data on this very interesting prominence-corona association. Other types of ascending prominences may also cause coronal transients, and these prominences may, or may not, be directly flare-related. We may therefore divide the coronal transients into two groups, those that are clearly flare-associated, and those that are not. In the case of the spray-produced transient it is probably the flare-produced shock-wave that is responsible for the sequence of events observed in the corona. The ascending material, responsible for a transient following a disparition brusque of a quiescent prominence, is accelerated in a way very different from spray material. Such ascending prominences start out very slowly and reach higher velocities only after considerable time, as the energy of the pre-existing, non-potential, magnetic field has been used to accelerate the prominence plasma. Again, the activation of the prominence may have been due to a flare (Chapter VI, 2.2), but it is probably the shock wave produced by the steadily accelerated ascending material that is responsible for the coronal response, the transient. Garcia *et al.* (1971) report cases where a large part of the white-light corona disappeared after the disparition brusque of a quiescent prominence. If there is this causal relationship between the disparition brusque and the coronal transient, the redistribution of magnetic fields during and after the disparition brusque phase may have a profound influence on the stability of coronal matter above the prominence. Coronal transients of this group have not been discussed extensively in the literature, even though some spectacular events have been recorded. Wild *et al.* (1968) draw attention to type II and moving type IV radio emission following a disparition brusque of a quiescent prominence (see also Giovanelli and Roberts, 1958; Warwick, 1965; Wild, 1970.

On the other hand, the clearly flare-associated transients have been devoted considerable attention, partly because of their spectacular nature. It is well established (McCabe and Fisher, 1970; Riddle, 1970; Dulk and Altschuler, 1971; Hansen *et al.*, 1971) that following spray-producing flares we may observe several phenomena in the corona, e.g. type II and type IV radio bursts, coronal whips – sudden changes in a previously existing arch structure seen in the Fe XIV, 5303 Å line (Dunn, 1960) – and white-light coronal transients. These phenomena may be considered different manifestations of the same disturbance (Smerd and Dulk, 1971; Kopp, 1972; Newkirk, 1971; see also Wild, 1969; Stewart and Sheridan, 1970). Kopp traces the sequence of events to a shock which is produced by the flare and, as it moves out, blows off part of the chromosphere. This blown-off material is interpreted as the spray and acts as the driver gas in the corona where the radio bursts follow.

Newkirk visualizes the sequence of events in the following way. As the flare is triggered, an MHD shock is formed that generates the spray prominence. The field lines readjust after the release of the magnetic energy (the source of the flare) and this readjustment takes place at the Alfvén speed and may be observed as coronal whips The shock wave expands out in the corona where it may produce type II and type IV radio bursts, as well as high-energy particles and X-ray emission. If the shock wave is reflected or refracted back to the chromosphere (along appropriate field lines) it may hit a distant prominence and cause a destabilization of the prominence (Chapter VI.2.2). If the shock wave escapes into interplanetary space, the velocity will be about 1000 km s^{-1} in the outer corona. Beyond this point the shock wave gradually decelerates, and by the time it reaches the vicinity of the Earth, its speed is about 700 km s^{-1}.

It is interesting to note that some sprays may perhaps be the only type of prominence that has a direct influence on terrestrial conditions.

Newkirk's assertions seem to be supported by a number of observations of sprays and moving type IV radio bursts, even though it may be difficult, in many cases, to show clearly the link between ascending prominences and the coronal "radio-plasma". Figure VII.2 shows the impressive ascending prominence of August 12, 1972. It could be photographed to heights in excess of 700000 km above the Sun. During this time one facet of a coronal transient was observed, i.e., the white-light corona changed in a manner suggesting the rise of closed helmet arch structures. Following the observation of the ascending prominence another facet of the coronal transient appeared, i.e., a type IV radio burst was observed to move out to a distance of about 4 solar radii (Riddle, 1973). Figure VII.3, which shows the relationship between the distance-time plot for the ascending prominence and for the radio event, strongly indicates that part of the spray material may have been accelerated to great heights where the plasma was responsible for the radio emission. This interpretation seems even more likely in view of observations with white-light coronagraphs (Sheridan *et al.*, 1972). Tousey and Koomen (1973), using the U.S. Naval Research Laboratory coronagraph onboard the OSO-7 spacecraft, have recorded clouds traveling out through the corona to distances of $10R_{\odot}$ following the observation of ascending prominences. Several varieties of coronal transients, as observed in the green coronal line (Fe XIV, 5303 Å), have been discussed by Dunn (1971) and DeMastus *et al.* (1973).

7.3. Miscellaneous

There are several other prominence-corona interactions, some on long time scales, than discussed in this chapter so far which, even though we shall mention them only briefly, may point to associations of great importance for our understanding of coronal physics.

7.3.1. THE SAN MANUEL EFFECT

A possible exchange of matter between quiescent prominences and the corona is revealed by a study of strong emission lines from the upper part of such prominences.

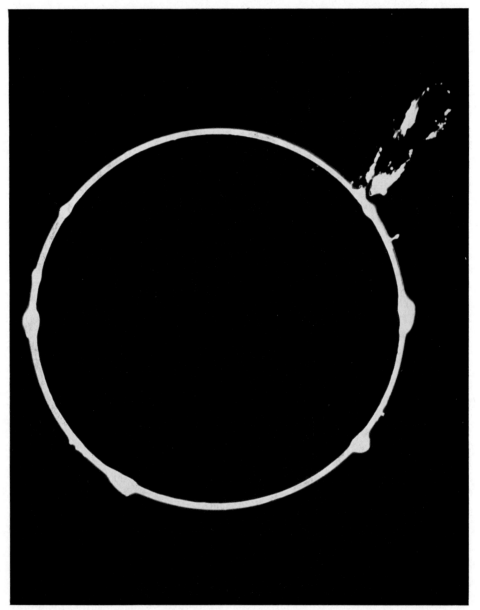

Fig. VII.2. Ascending prominence of August 12, 1972. Mauna Loa station of the HAO.

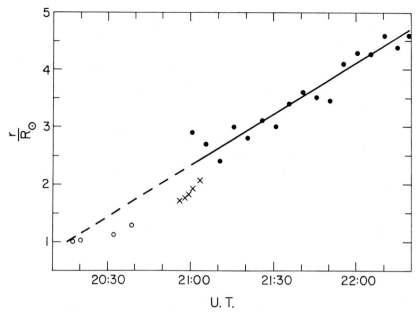

Fig. VII.3. Radial distance, $r/R_\odot$, of the moving type IV radio event of August 12, 1972, as a function of time, dots (Riddle, 1973), and positions of ascending prominence material as seen in Hα. Open circles Boulder flare patrol, courtesy Natl. Oceanic and Atmosph. Adm.; crosses Mauna Loa station of the HAO.

Evershed (1927, 1929, 1945) deduced a rotational velocity of prominences from measurements of the Doppler displacement of Ca II, K and H lines, and found that the velocity was significantly greater than the angular velocity inferred from the times of limb passages of the prominences. He suggested that a west wind in the solar atmosphere might account for the discrepancy.

The 'west wind hypothesis' has been revived in a modern form by Livingston (1971). A closer inspection of the emission line profiles reveals a peculiar shape. When the spectra are taken with the slit normal to the solar limb crossing the prominence, the top of the emission line, from the upper edge of the filament, shows an abrupt line shift. The analogy with a picture of smoke blown from a chimney is striking, and Livingston has called the phenomenon the San Manuel effect, after a smelter stack near Tucson, Arizona. The spectroscopic features have been analyzed in more detail by Engvold and Livingston (1971), who suggest that the fuzziness of the upper part of the emission lines indicates an exchange of matter between the prominence and the corona.

7.3.2. STREAMER-PROMINENCE INDICATOR

Axisa et al. (1971), have shown that the radio sources responsible for the sun's slowly varying component at 169 MHz are linked to the presence of quiescent prominences. These filaments are found at the base of streamers in which – much higher up in the corona ($h \approx 270\,000$ km) – the radio source is situated. The authors consider the radio sources as the radio counterparts of streamers, which means that there is a fundamental association between a streamer and the quiescent prominence found at its base (see Section 1.3).

REFERENCES

Alfvén, H. and Carlqvist, P.: 1967, *Solar Phys.* **1**, 220.
Ananthakrishnan, R. and Madhavan Nayar, P.: 1953, *Kodiakanal Obs. Bull.* No. 137, 194.
Anderson, G. F.: 1966, Ph.D. Thesis, Univ. of Colorado.
Anzer, U.: 1969, *Solar Phys.* **8**, 37.
Anzer, U. and Tandberg-Hanssen, E.: 1970, *Solar Phys.* **11**, 61.
Athay, R. G.: 1966, *Astrophys. J.* **146**, 223.
Athay, R. G.: 1972, *The Solar Chromosphere and Corona,* D. Reidel Publ. Co., Dordrecht, Holland.
Athay, R. G. and Johnson, H. R.: 1960, *Astrophys. J.* **131**, 413.
Athay, R. G. and Moreton, G. E.: 1961, *Astrophys. J.* **133**, 935.
Athay, R. G. and Skumanich, A.: 1967, *Ann. Astrophys.* **30**, 669.
Avrett, E. H. and Loeser, R.: 1963, *J. Quant. Spectr. Radiative Transfer* **3**, 201.
Axisa, F., Avignon, Y., Martres, M. J., Pick, M., and Simon, P.: 1971, *Solar Phys.* **19**, 110.
Babcock, H. D. and Babcock, H. W.: 1955, *Astrophys. J.* **121**, 349.
Balanovsky, J. and Perepelkin, E.: 1928, *Monthly Notices Roy. Astron. Soc.* **88**, 740.
Becker, U.: 1957, *Z. Astrophys.* **42**, 85.
Becker, U.: 1958, *Z. Astrophys.* **44**, 243.
Beckers, J. M.: 1966, *Anacapri Symp.* p. 80.
Beckers, J. H.: 1968a, *Solar Phys.* **3**, 367.
Beckers, J. H.: 1968b, *Solar Phys.* **5**, 15.
Bernstein, I. B., Frieman, E. K., Kruskal, M. D., and Kulsrud, R. M.: 1958, *Proc. Roy. Soc.* **A244**, 17.
Bhatnager, P. L., Krook, M., and Menzel, D. H.: 1951, 'Dynamics of Ionized Media', Rep. Conf. on Dynamics of Ionized Media, Univ. College, London.
Billings, D. E.: 1966, *A Guide to the Solar Corona,* Academic Press, New York.
Bishop, A. S.: 1966, AEC Res. Dev. Rep. MATT-412, Princeton Univ.
Bjerke, K.: 1961, *Arkiv Astron.* **2**, 145.
Bratiichuk, M. V.: 1961, *Publ. Kiev Obs.* **9**, 11.
Bray, R. J. and Loughhead, R. E.: 1964, *Sunspots,* John Wiley and Sons, Inc., New York.
Brown, A.: 1958, *Astrophys. J.* **128**, 646.
Bruzek, A.: 1951, *Z. Astrophys.* **13**, 277.
Bruzek, A.: 1952, *Z. Astrophys.* **31**, 99.
Bruzek, A.: 1957, *Z. Astrophys.* **42**, 76.
Bruzek, A.: 1958, *Z. Astrophys.* **44**, 183.
Bruzek, A.: 1964a, *Astrophys. J.* **140**, 746.
Bruzek, A.: 1964b, *J. Geophys. Res.* **69**, 2386.
Bruzek, A.: 1967, *Solar Phys.* **2**, 451.
Bruzek, A.: 1968, in K. O. Kiepenheuer (ed.), 'Structure and Development of Solar Active Regions', *IAU Symp.* **35**, 293.
Bruzek, A.: 1969a, *Solar Phys.* **8**, 29.
Bruzek, A.: 1969b, in C. de Jager and Z. Švestka (eds.), *Solar Flares and Space Research,* North-Holland, Amsterdam, p. 61.
Bruzek, A. and DeMastus, H. L.: 1970, *Solar Phys.* **12**, 447.
Budden, K. G.: 1961, *Radio Waves in the Ionosphere,* Cambridge Univ. Press, London.
Buhl, D. and Tlamicha, A.: 1970, *Astron. Astrophys.* **5**, 102.
Bumba, V. and Howard, R.: 1965, *Astrophys. J.* **141**, 1492.
Bumba, V. and Kleczek, J.: 1961, *Observatory* **81**, 141.
Burns, J.: 1970, Ph.D. Thesis, Univ. of Hawaii.
Carlqvist, P.: 1969, *Solar Phys.* **7**, 220.
Celsius, A.: 1735, *Acta Lit. Sci. Sueciae* **4**, 48.
Chandrasekhar, S.: 1960, *Radiative Transfer,* Dover, New York.

Chandrasekhar, S. and Breen, F. H.: 1946, *Astrophys. J.* **104**, 430.
Chandrasekhar, S. and Kendall, P. C.: 1957, *Astrophys. J.* **126**, 457.
Christiansen, W. N., Mathewson, D. S., Pawsey, J. L., Smerd, S. F., Boischot, A., Denisse, J. F., Simon, P., Kakinuma, T., Dodson-Prince, H., and Firor, J.: 1960, *Ann. Astrophys.* **23**, 75.
Chu, C. K. and Grad, H.: 1965, in R. J. Seeger and G. Temple (eds.), *Magnetofluid Dynamics in Research Frontiers in Fluid Dynamics,* Interscience, New York.
Clark, C. D. and Park, W. M.: 1968, *Nature* **219**, 922.
Collett, E.: 1968, *Am. J. Phys.* **36**, 713.
Conway, M. T.: 1952, *Proc. Roy. Irish Acad.* **54A**, 311.
Coppi, B.: 1964, *Ann. Phys.* **30**, 178.
Correll, M. and Roberts, W. O.: 1958, *Astrophys. J.* **127**, 726.
Correll, M., Hazen, M., and Bahng, J.: 1956, *Astrophys. J.* **124**, 597.
Covington, A. E.: 1954, *J. Geophys. Res.* **59**, 163.
Cowling, T. G.: 1957, *Magnetohydrodynamics,* Interscience Press, New York.
Cox, D. P. and Tucker, W. H.: 1969, *Astrophys. J.* **157**, 1157.
Davidson, C. R. and Stratton, F. J. M.: 1927, *Mem. Roy. Astron. Soc.* **64**, Part 4, Sec. 4, 105.
d'Azambuja, L. and d'Azambuja, M.: 1948, *Ann. Obs. Paris-Meudon* **6**, 7.
de Jager, C.: 1952, *Rech. Astron. Obs. Utrecht* **13**, Part I.
de Jager, C.: 1959, *Handbuch der Physik* **52**, 80.
de la Rue, W.: 1868, *Monthly Notices Roy. Astron. Soc.* **29**, 73.
DeMastus, H. L., Wagner, W. J., and Robinson, R. D.: 1974, *Solar Phys.* in press.
Deslandres, H.: 1897, *Compt. Rend. Acad. Sci.* **124**, 171.
Deslandres, H.: 1910, *Ann. Obs. Paris-Meudon,* **4**, I, 1.
Dodson-Prince, H. W.: 1949, *Astrophys. J.* **110**, 382.
Dodson-Prince, H. W.: 1961, *Proc. Nat. Acad. Sci.* **47**, 901.
Dodson-Prince, H. W. and Hedeman, E. R.: 1964, *AAS-NASA Symp. on the Physics of Solar Flares 1963,* p. 15.
Doherty, L. R. and Menzel, D. H.: 1965, *Astrophys. J.* **141**, 251.
Drago, F. C. and Felli, M.: 1970, *Solar Phys.* **14**, 171.
Dulk, G. A. and Altschuler, M. D.: 1971, *Solar Phys.* **20**, 438.
Dungey, J. W.: 1953, *Monthly Notices Roy. Astron. Soc.* **113**, 180.
Dungey, J. W.: 1958, *Cosmic Electrodynamics,* Cambridge Univ. Press, London.
Dunn, R.: 1960, Ph.D. Thesis, Harvard Univ.
Dunn, R.: 1965, Sacramento Peak Obs. Contr. #87.
Dunn, R.: 1970, *AIAA Bull.* **7**, 564.
Dunn, R.: 1971, in C. Macris (ed.), *Physics of the Solar Corona,* D. Reidel Publ. Co., Dordrecht, Holland, p. 114.
Dyson, F.: 1930, *Monthly Notices Roy. Astron. Soc.* **91**, 239.
Efanov, V., Moiseev, I., and Severny, A. B.: 1968, in K. O. Kiepenheuer (ed.) 'Structure and Development of Solar Active Regions', *IAU Symp.* **35**, 588.
Ellison, M. A.: 1944, *Monthly Notices Roy. Astron. Soc.* **104**, 22.
Engvold, O.: 1971, 'Physics of Solar Prominences', Coll. Anacapri German Solar Obs.
Engvold, O. and Livingston, W.: 1971, *Solar Phys.* **20**, 375.
Evershed, J.: 1927, *Monthly Notices Roy. Astron. Soc.* **88**, 126.
Evershed, J.: 1929, *Monthly Notices Roy. Astron. Soc.* **89**, 250.
Evershed, J.: 1945, *Monthly Notices Roy. Astron. Soc.* **105**, 204.
Fermi, E.: 1949, *Phys. Rev.* **75**, 1169.
Fermi, E.: 1954, *Astrophys. J.* **119**, 1.
Ferraro, V. C. A. and Plumpton, C.: 1966, *An Introduction to Magneto-Fluid Mechanics,* 2nd ed., Clarendon Press, Oxford.
Fichtel, C. E. and McDonald, F. B.: 1967, *Ann. Rev. Astron. Astrophys.* **5**, 351.
Field, G. B.: 1965, *Astrophys. J.* **142**, 531.
Firor, J.: 1959, in R. N. Bracewell (ed.), 'Paris Symposium on Radio Astronomy', *IAU Symp.* **9**, 136.
Fredga, K.: 1969, *Solar Phys.* **9**, 358.
Garcia, C., Hansen, R., Hull, H., Lacey, L., and Lee, R.: 1971, *Bull. Am. Astron. Soc.* **3**, 261.
Giovanelli, R. G.: 1946, *Nature* **158**, 81.
Giovanelli, R. G.: 1947, *Monthly Notices Roy. Astron. Soc.* **107**, 338.
Giovanelli, R. G.: 1948, *Monthly Notices Roy. Astron. Soc.* **108**, 163.
Giovanelli, R. G.: 1967, *Australian J. Phys.* **20**, 81.
Giovanelli, R. G. and McCabe, M.: 1958, *Australian J. Phys.* **11**, 191.
Giovanelli, R. G. and Roberts, J. A.: 1958, *Australian J. Phys.* **11**, 353.

Godovnikov, N. V. and Smirnova, E. P.: 1965, *Izv. Krymsk. Astrofiz. Observ.* **33**, 86.
Gold, T.: 1964, *AAS-NASA Symp. on the Physics of Solar Flares 1963*, p. 389.
Gold, T.: 1968, *Nobel Symp.* **9**, 206.
Gold, T.: 1971, private communication.
Gold, T. and Hoyle, F.: 1960, *Monthly Notices Roy. Astron. Soc.* **120**, 89.
Goldberg, L.: 1939, *Astrophys. J.* **89**, 673.
Goldsmith, D. W.: 1971, *Solar Phys.* **19**, 86.
Gopasyuk, S. I.: 1960, *Izv. Krymsk. Astrofiz. Observ.* **23**, 331.
Gopasyuk, S. I. and Ogir, M. B.: 1963, *Izv. Krymsk. Astrofiz. Observ.* **30**, 185.
Grad, H. and Rubin, H.: 1958, *Proc. 2nd Int. Conf. Peaceful Uses of Atomic Energy* **31**, 190.
Grant, R.: 1852, *History of Physical Astronomy*, Robert Baldwin, London.
Green, R. M.: 1965, in R. Lüst (ed.), 'Stellar and Solar Magnetic Fields', *IAU Symp.* **22**, 398.
Griem, H. R.: 1960, *Astrophys. J.* **132**, 883.
Griem, H. R.: 1962, *Astrophys. J.* **136**, 422.
Grotrian, W.: 1931, *Z. Astrophys.* **2**, 106.
Hale, G. E.: 1929, *Astrophys. J.* **70**, 265.
Hale, G. E.: 1930, *Astrophys. J.* **71**, 73.
Hale, G. E.: 1931a, *Astrophys. J.* **73**, 379.
Hale, G. E.: 1931b, *Astrophys. J.* **74**, 214.
Hale, G. E. and Ellerman, F.: 1903, *Publ. Yerkes Obs.* **3**, Part I, 3.
Hansen, R. T., Garcia, C. J., Grognard, R. J.-M., and Sheridan, K. V.: 1971, *Proc. Astron. Soc. Australia* **2**, 57.
Hansen, S. F., Hansen, R. T., and Garcia, C. J.: 1972, *Solar Phys.* **26**, 202.
Harris, D. L.: 1948, *Astrophys. J.* **108**, 112.
Harvey, J. W.: 1969, Ph.D. Thesis, Univ. of Colorado.
Harvey, J. and Tandberg-Hanssen, E.: 1968, *Solar Phys.* **3**, 316.
Haselgrove, J.: *Rep. Conf. on the Physics of the Ionosphere*, p. 355.
Haselgrove, J.: 1957, *Proc. Phys. Soc.* **B.70**, 653.
Haupt, H.: 1955, *Mitt. Obs. Kanzelhöhe*, No. 6.
Hayakawa, S. J., Nishimura, H., Obayashi, H., and Sato, H.: 1964, *Prog. Theor. Phys. Suppl.* **30**, 86.
Hildner, E.: 1971, Ph.D. Thesis, Univ. of Colorado.
Hildner, E.: 1972, private communication.
Hirayama, T.: 1963, *Publ. Astron. Soc. Japan* **15**, 122.
Hirayama, T.: 1964, *Publ. Astron. Soc. Japan* **16**, 104.
Hirayama, T.: 1971a, *Solar Phys.* **17**, 50.
Hirayama, T.: 1971b, Anacapri Coll. on 'The Physics of Prominences', German Solar Obs.
Hjerting, F.: 1938, *Astrophys. J.* **88**, 508.
Howard, R.: 1959, *Astrophys. J.* **130**, 193.
Huggins, W.: 1869, *Proc. Roy. Soc.* (*London*) **17**, 302.
Hunter, J. H.: 1966, *Icarus* **5**, 321.
Hunter, J. H.: 1970, *Astrophys. J.* **161**, 451.
Hyder, C. L.: 1964a, Ph.D. Thesis, Univ. of Colorado.
Hyder, C. L.: 1964b, *Astrophys. J.* **140**, 817.
Hyder, C. L.: 1966, *Z. Astrophys.* **63**, 78.
Hyder, C. L.: 1967a, *Solar Phys.* **2**, 49.
Hyder, C. L.: 1967b, *Solar Phys.* **2**, 267.
Hyder, C. L.: 1968, *Nobel Symp.* **9**, 57.
Idlis, G. M., Karimov, M. G., Delone, A. B., and Obashev, S. O.: 1956, *Izv. Ap. Inst. Akad. Nauk Kazakhskoi SSR* **11**, 71.
Inglis, D. R. and Teller, E.: 1939, *Astrophys. J.* **90**, 439.
Ioshpa, B. A.: 1962, *Geomag. Aeron.* **2**, 149.
Ioshpa, B. A.: 1963, *Geomag. Aeron.* **3**, 903.
Ioshpa, B. A.: 1965, *Astron. Zh.* **42**, 754.
Ioshpa, B. A.: 1968, Results of Research on the Intl. Geophys. Projects. Solar Activity No. 3, Nauka, Moscow, p. 44.
Ivanov-Kholodny, G. S.: 1955, *Izv. Krymsk. Astrofiz. Observ.* **13**, 112.
Ivanov-Kholodny, G. S.: 1958, *Izv. Krymsk. Astrofiz. Observ.* **18**, 109.
Ivanov-Kholodny, G. S.: 1959, *Astron. Zh.* **36**, 589. (*Soviet Astron.* **3**, 578, 1960.)
Janssen, P. J.: 1868, *Compt. Rend. Acad. Sci.* **67**, 838.
Jefferies, J. T.: 1953, *Australian J. Phys.* **6**, 27.
Jefferies, J. T.: 1968, *Spectral Line Formation*, Blaisdell, Waltham.

Jefferies, J. T. and Orrall, F. Q.: 1958, *Astrophys. J.* **127**, 714.
Jefferies, J. T. and Orrall, F. Q.: 1961a, *Astrophys. J.* **133**, 946.
Jefferies, J. T. and Orrall, F. Q.: 1961b, *Astrophys. J.* **133**, 963.
Jefferies, J. T. and Orrall, F. Q.: 1961c, *Astrophys. J.* **134**, 747.
Jefferies, J. T. and Orrall, F. Q.: 1962, *Astrophys. J.* **135**, 109.
Jefferies, J. T. and Orrall, F. Q.: 1963, *Astrophys. J.* **137**, 1232.
Jefferies, J. T. and Orrall, F. Q.: 1964, *AAS-NASA Symp. on The Physics of Solar Flares 1963*.
Jefferies, J. T. and Orrall, F. Q.: 1965, *Astrophys. J.* **141**, 519.
Jefferies, J. T. and Thomas, R. N.: 1958, *Astrophys. J.* **127**, 667.
Jefferies, J. T. and Thomas, R. N.: 1960, *Astrophys. J.* **131**, 695.
Jeffrey, A. and Tanuiti, T.: 1966, in A. Jeffrey and T. Tanuiti (eds.), *Magnetohydrodynamic Stability and Thermonuclear Containment*, Academic Press, New York, p. 1.
Jensen, E.: 1959, *Astrophys. Norv.* **6**, 93.
Jette, A. D. and Sreenivasan, S. R.: 1969 *Phys. Fluids* **12**, 2544.
Kawaguchi, I.: 1964, *Publ. Astron. Soc. Japan* **16**, 86.
Kawaguchi, I.: 1965, *Publ. Astron. Soc. Japan* **17**, 367.
Kawaguchi, F.: 1967, *Solar Phys.* **1**, 420.
Khangildin, U.: 1964, *Astron. Zh.* **41**, 302.
Kiepenheuer, K. O.: 1938, *Z. Astrophys.* **15**, 53.
Kiepenheuer, K. O.: 1953a, in G. Kuiper (ed.), *The Sun*, Ch. 4, p. 327, Chicago Univ. Press.
Kiepenheuer, K. O.: 1953b, *Accad. Naz. Lincei, Conv. Volta* **11**, 148.
Kiepenheuer, K. O.: 1959, *Z. Astrophys.* **48**, 290.
Kiepenheuer, K. O.: 1968, *Nobel Symp.* **9**, 123.
Kippenhahn, R. and Schlüter, A.: 1957, *Z. Astrophys.* **43**, 36.
Kirshner, R. P. and Noyes, R. W.: 1971, *Solar Phys.* **20**, 428.
Kleczek, J.: 1957, *Bull. Astron. Inst. Czech.* **8**, 120.
Kleczek, J.: 1958, *Bull. Astron. Inst. Czech.* **9**, 115.
Kleczek, J.: 1962, *Astron. J.* **67**, 275.
Kleczek, J.: 1963, in J. Evans (ed.), *The Solar Corona*, Academic Press, New York, p. 151.
Kleczek, J.: 1964, *AAS-NASA Symp. on the Physics of Solar Flares 1963*, p. 77.
Kleczek, J.: 1968, in K. O. Kiepenheuer (ed.), 'Structure and Development in Solar Active Regions', *IAU Symp.* **35**, 280.
Kleczek, J.: 1969, *Solar Phys.* **7**, 238.
Kleczek, J. and Hansen, R. T.: 1962, *Publ. Astron. Soc. Pacific* **74**, 507.
Kleczek, J. and Kuperus, M.: 1969, *Solar Phys.* **6**, 72.
Kopp, R. A.: 1972, *Asilomar Conference on Solar Wind 1971*, p. 252.
Kundu, M. R.: 1959, *Ann. Astrophys.* **22**, 1.
Kundu, M. R.: 1965, *Solar Radio Astronomy*, Interscience Press, New York.
Kundu, M. R., Erickson, W. C., and Jackson, P. D.: 1970, *Solar Phys.* **14**, 394.
Kuperus, M.: 1965, *Res. Astron. Obs. Utrecht* **17**, 1.
Kuperus, M. and Athay, R. G.: 1967, *Solar Phys.* **1**, 361.
Kuperus, M. and Tandberg-Hannsen, E.: 1967, *Solar Phys.* **2**, 39.
Lange, I. and Forbush, S. E.: 1942. *Terrestr. Magn. Atmospheric Electr.* **47**, 185.
Larmore, L.: 1953, *Astrophys. J.* **118**, 436.
Ledoux, P. and Walraven, T.: 1958, *Handbuch der Physik* **51**, 353.
Leroy, J. L. and Servajean, R.: 1966, *Ann. Astrophys.* **29**, 263.
Lindsay, R. B.: 1960, *Mechanical Radiation*, McGraw-Hill, New York.
Linhart, J. G.: 1960, *Plasma Physics*, North-Holland Publ. Co., Amsterdam, p. 186
Lippincott, S. L.: 1955, *Ann. Astrophys.* **18**, 113.
Livingston, W.: 1971, *Solar Phys.* **19**, 379.
Lockyer, J. N.: 1868a, *Compt. Rend. Acad. Sci.* **67**, 836.
Lockyer, J. N.: 1868b, *Compt. Rend. Acad. Sci.* **67**, 949.
Lockyer, W. J. S.: 1903, *Monthly Notices Roy. Astron. Soc.* **63**, 446.
Lockyer, W. J. S.: 1922, *Monthly Notices Roy. Astron. Soc.* **82**, 323.
Lockyer, W. J. S.: 1931, *Monthly Notices Roy. Astron. Soc.* **91**. 797
Lodén, K.: 1958, *Arkiv. Astron.* **2**, 153.
Lüst, R. and Schlüter, A.: 1954, *Z. Astrophys.* **34**, 263.
Lüst, R. and Zirin, H.: 1960, *Z. Astrophys.* **49**, 8.
Lyot, B.: 1936, *Compt. Rend. Acad. Sci.* **202**, 392.
Lyot, B.: 1937, *Bull. Soc. Astron. France* **51**, 203.
Lyot, B.: 1939, *Monthly Notices Roy. Astron. Soc.* **99**. 580.

Malville, J.: 1961, Ph.D. Thesis, Univ. of Colorado.
Malville, J.: 1968 *Solar Phys.* **5**, 236.
Malville, J.: 1969, private communication.
Malville, J. M. and Tandberg-Hanssen, E.: 1969, *Solar Phys.* **6**, 278.
Martres, M.-J.: 1956, *L'Astronomie* **70**, 401.
Martres, M.-J., Michard, R., and Soru-Iscovici, J.: 1966, *Ann. Astrophys.* **29**, 245.
McCabe, M.: 1971, *Solar Phys.* **19**, 451.
McCabe, M. and Fisher, R. R.: 1970, *Solar Phys.* **14**, 212.
McMath, R. R. and Pettit, E.: 1937, *Astrophys. J.* **85**, 279.
McMath, R. R. and Pettit, E.: 1938, *Astrophys. J.* **88**, 244.
Menzel, D. H. and Evans, J. W.: 1953, *Accad. Naz. Lincei. Convegno Volta* **11**, 119.
Menzel, D. H., Smith, E. v. P., DeMastus, H., Ramsey, H., Schnable, G., and Lawrence, R.: 1956, *Astron. J.* **61**, 186.
Meyer, F.: 1968, in K. O. Kiepenheuer (ed.), 'Structure and Development of Solar Active Regions', *IAU Symp.* **35**, 485.
Meyer, F. and Schmidt, H. U.: 1968, *Z. Angew. Math. Mech.* **48**, 218.
Michard, R.: 1961, *Compt. Rend. Acad. Sci.* **253**, 2857.
Mitchell, S. A.: 1935, *Eclipses of the Sun,* Columbia Univ. Press, New York.
Miyamoto, S.: 1947, *Mem. College Sci. Kyoto Univ.* **25**, 31.
Molodensky, M. M.: 1966, *Astron. Zh.* **43**, 727.
Moreton, G E.: 1960, *Astron. J.* **65**, 494.
Moreton, G. E.: 1965, in R. Lüst (ed.), 'Stellar and Solar Magnetic Fields,' *IAU Symp.* **22**, 371.
Moreton, G. E. and Ramsey, H.: 1960, *Publ. Astron. Soc. Pacific* **72**, 357.
Moroshenko, N. N.: 1963, *Bull. Main Astrophys. Observ. Kiev* **5**, No. 1, 93.
Mott, N. F. and Massey, H. S. W.: 1949, *The Theory of Atomic Collisions,* Clarendon Univ. Press, Oxford.
Nakagawa, Y.: 1970, *Solar Phys.* **12**, 419.
Nakagawa, Y. and Hyder, C. L.: 1969, Environmental Res. Paper #320, AFCRL-70-0273, Office of Aerospace Res., USAF.
Nakagawa, Y. and Malville, J. M.: 1969, *Solar Phys.* **9**, 102.
Nakagawa, Y., Raadu, M. A., Billings, D. E., and McNamara, E.: 1971, *Solar Phys.* **19**, 72.
Neupert, W. M.: 1969, *Ann. Rev. Astron. Astrophys.* **7**, 121.
Newkirk, G. N. Jr.: 1957, *Ann. Astrophys.* **20**, 127.
Newkirk, G. N. Jr.: 1967, *Ann. Rev. Astron. Astrophys.* **5**, 213.
Newkirk, G. N. Jr.: 1972, *Asilomar Conference on Solar Wind, 1971,* p. 11.
Newton, H. W.: 1934, *Monthly Notices Roy. Astron. Soc.* **94**, 472.
Newton, H. W.: 1935, *Monthly Notices Roy. Astron. Soc.* **95**, 650.
Noyes, R. W.: 1972, private Communication.
Obridko, V. N.: 1965, *Soviet Astron.* **9**, 398.
Obridko, V. N.: 1968, Results of Researches on the Intl. Geophys. Projects, Solar Activity No. 3, Nauka, Moscow, p. 71.
Øhman, Y.: 1971a, *Trans. IAU* **XIVB**, 105.
Øhman, Y.: 1971b, 'Physics of Prominences', Coll. Anacapri, German Solar Obs.
Øhman, Y.: 1972 *Solar Phys.* **23**, 134.
Øhman, Y. and Øhman, N.: 1953, *Observatory* **73**, 203.
Øhman, Y., Lindgren, L., and Lindgren, U.: 1962, *Arkiv. Astron.* **3**, 121.
Øhman, Y., Stiber, G., and Kusoffsky, U.: 1967, *Solar Phys.* **1**, 60.
Olson, C. A. and Lykoudis, P. S.: 1967, *Astrophys. J.* **150**, 303.
Orrall, F. Q. and Athay, R. G.: 1957, *Astron. J.* **62**, 28.
Orrall, F. Q. and Zirker, J. B.: 1961, *Astrophys. J.* **134**, 72.
Oster, L. and Sofia, S.: 1966, *Astrophys. J.* **143**, 944.
Osterbrock, D. E.: 1961, *Astrophys. J.* **134**, 347.
Pannekoek, A. and Doorn, N. W.: 1930, *Verh. Akad. Wetensch. (Amsterdam)* **14**, No. 2.
Parker, E. N.: 1953, *Astrophys. J.* **117**, 431.
Parker, E. N.: 1958, *Phys. Rev.* **109**, 1328.
Parker, E. N.: 1963a, *Astrophys. J. Suppl.* **8**, 177.
Parker, E. N.: 1963b, *Astrophys. J.* **138**, 552.
Parker, E. N.: 1964, *Astrophys. J.* **140**, 1170.
Pawsey, J. L. and Bracewell, R. N.: 1955, *Radio Astronomy,* Clarendon Press, Oxford.
Petschek, H. E.: 1964, *AAS-NASA Symp. on the Physics of Solar Flares 1963,* p. 425.
Pettit, E.: 1919, *Astrophys. J.* **50**, 206.

Pettit, E.: 1925, *Publ. Yerkes Obs.* **3**, IV, 205.
Pettit, E.: 1932, *Astrophys. J.* **76**, 9.
Pettit, E.: 1936, *Astrophys. J.* **84**, 319.
Pettit, E.: 1938, *Publ. Astron. Soc. Pacific* **50**, 168.
Pettit, E.: 1943, *Astrophys. J.* **98**, 6.
Pettit, E.: 1950, *Publ. Astron. Soc. Pacific* **62**, 144.
Pikel'ner, S. B.: 1971, *Solar Phys.* **17**, 44.
Pikel'ner, S. B. and Livshitz, M. A.: 1964, *Astron. Zh.* **41**, 1007.
Pneuman, G. W.: 1972, *Astrophys. J.* **177**, 793.
Pneuman, G. W. and Kopp, R. A.: 1970, *Solar Phys.* **13**, 176.
Poland, A. and Anzer, U.: 1971, *Solar Phys.* **19**, 401.
Poland, A., Skumanich, A., Athay, R. G., and Tandberg-Hanssen, E.: 1971, *Solar Phys.* **18**, 391.
Pottasch, S. R.: 1965, *Bull. Astron. Inst. Neth.* **18**, 7.
Priest, E. R.: 1972a, *Quart. J. Mech. Appl. Math.* **25**, pt. 3.
Priest, E. R.: 1972b, *Monthly Notices Roy. Astron. Soc.* **159**, 389.
Raadu, M.: 1971, private communication.
Raadu, M. A. and Nakagawa, Y.: 1971, *Solar Phys.* **20**, 64.
Rachovsky, D. N.: 1961, *Izv. Krymsk. Astrofiz. Observ.* **26**, 63.
Rachovsky, D. N.: 1963, *Izv. Krymsk. Astrofiz. Observ.* **30**, 267.
Rachovsky, D. N.: 1967, *Izv. Krymsk. Astrofiz. Observ.* **36**, 51.
Raju, P. K.: 1966, Ph.D. Thesis, Yale University.
Raju, P. K.: 1968, *Monthly Notices Roy. Astron. Soc.* **139**, 479.
Ramsey, H. and Smith, S. F.: 1966, *Astron. J.* **71**, 197.
Ramsey, H. E., Smith, S. F., and Angle, K. L.: 1968, Final Rep. High Resol. Solar Phot. LSMC 681495. Lockheed Palo Alto Res. Lab.
Reber, E. E.: 1971, *Solar Phys.* **16**, 75.
Redman, R. O. and Zanstra, H.: 1952, *Proc. Nederl. Akad. Wetensch.* **55B**, 598.
Richardson, R. S.: 1950, *Astrophys. J.* **111**, 572.
Riddle, A. C.: 1970, *Solar Phys.* **13**, 448.
Riddle, A. C.: 1973, Univ. of Colo. Radio Astr. Obs. Rep. SN-2.
Rigutti, M. and Russo, D.: 1961, *Accad. Naz. Lincei,* Series 8, **30**, 487.
Roberts, W. O. and Billings, D. E.: 1955a, *Astron. J.* **60**, 176.
Roberts, W. O. and Billings, D. E.: 1955b, HAO Solar Research Memo #33.
Rompolt, B.: 1971, 'Physics of Prominences', Coll. Anacapri, German Solar Obs.
Rosseland, S.: 1926, *Astrophys. J.* **63**, 342.
Rosseland, S.: 1936, *Astrophys. Norv.* **2**, 173.
Rosseland, S. and Tandberg-Hanssen, E.: 1957, *Astrophys. Norv.* **5**, 279.
Rosseland, S., Jensen, E., and Tandberg-Hanssen, E.: 1958, in B. Lehnert (ed.), 'Electromagnetic Phenomena in Cosmical Physics', *IAU Symp.* **6**, 150.
Rust, D.: 1966, Ph.D. Thesis, Univ. of Colorado.
Rust, D.: 1967, *Astrophys. J.* **150**, 313.
Rust, D.: 1968, in K. O. Kiepenheuer (ed.), 'Structure and Development of Solar Active Regions', *IAU Symp.* **35**, 77.
Rust, D. M. and Roy, J. R.: 1971, in R. Howard (ed.), 'Solar Magnetic Fields', *IAU Symp.* **43**, 569
Rust, D. and Smith, S. F.: 1969, AAS Pasadena Meeting.
Saito, K. and Hyder, C. L.: 1968, *Solar Phys.* **5**, 61.
Schatzman, E.: 1961, *Ann. Astrophys.* **24**, 251.
Schatzman, E.: 1963, *Ann. Astrophys.* **26**, 234.
Schlüter, A.: 1957a, *Z. Naturf.* **12a**, 822.
Schlüter, A.: 1957b, in H. C. van de Hulst (ed.), 'Radio Astronomy', *IAU Symp.* **4**, 356.
Schmidt, H. U.: 1964, *AAS-NASA Symp. on the Physics of Solar Flares 1963*, p. 107.
Schmidt, H. U.: 1965, *Mitt. Astr. Ges.* Sept. 19.
Schmidt, H. U.: 1966, Proc. Meet. Solar Magn. Fields & High Res. Spect., G. Barbera Firenze, p. 233.
Schmidt, H. U.: 1968, in K. O. Kiepenheuer (ed.), 'Structure and Development of Solar Active Regions,' *IAU Symp.* **35**, 95.
Schoolman, S. A.: 1969, Ph.D. Thesis, Univ. of Colorado.
Schwarzschild, K.: 1906, *Astron. Mitt. Gøttingen*, No. 13.
Secchi, A.: 1875–77, *Le Soleil*, Gauthier-Villars, Paris, Vols. 1 and 2.
Secchi, A.: 1868, *Compt. Rend. Acad. Sci.* **66**, 398.
Semel, M.: 1967, *Ann. Astrophys.* **30**, 513.
Severny, A. B.: 1950, *Dokl. Akad. Nauk.* 475.

Severny, A. B.: 1954, *Astron. Zh.* **31**, 131.
Severny, A. B.: 1959, *Solar Physics* (transl. by G. Yankovsky), Foreign Languages Pub. House, Moscow.
Severny, A. B.: 1964, *Ann. Rev. Astron. Astrophys.* **2**, 363.
Severny, A. B.: 1965, *IAU Trans.* **XIIA**, 755.
Severny, A. B. and Khoklova, V. L. 1953, *Izv. Krymsk. Astrofiz. Observ.* **10**, 9.
Shaposhnikova, E. F.: 1958, *Izv. Krymsk. Astrofiz. Observ.* **18**, 151.
Sheridan, K., Garcia, C., and Hansen, R.: 1972, AAS-Solar Phys. Meet. April, Univ. of Maryland.
Shin-Huei, E.: 1961, *Izv. Krymsk. Astrofiz. Observ.* **25**, 180.
Shklovsky, I. S.: 1960, *Cosmic Radio Waves*, Harvard Univ. Press, Cambridge, Mass.
Shklovsky, I. S.: 1965, *Physics of the Solar Corona*, Ch. 7.33 (Russian 2nd ed. Moscow 1962).
Shurcliff, W. A.: 1962, *Polarized Light*, Harvard Univ. Press, Cambridge.
Simon, G. W. and Leighton, R. B.: 1964, *Astrophys. J.* **140**, 1120.
Simon, M.: 1971, private communication.
Simon, M. and Wickstrøm, B. A.: 1971, *Solar Phys.* **20**, 122.
Slomin, Yu. M.: 1969, *Soviet Astron.* **13**, 450.
Smerd, S. F. and Dulk, G. A.: 1971, in R. Howard (ed.), 'Solar Magnetic Fields', *IAU Symp.* **43**, 616.
Smith, D.: 1970, *Solar Phys.* **15**, 202.
Smith, D. F. and Priest, E. R.: 1972, *Astrophys. J.* **176**, 487.
Smith, E. v. P.: 1968, *Nobel Symp.* **9**, 137.
Smith, E. v. P.: 1970, *Trans. IAU* **XIVA**, 96.
Smith, H. J. and Smith, E. v. P.: 1963, *Solar Flares*, Macmillan, New York.
Smith, S. F.: 1968, in K O. Kiepenheuer (ed.), 'Structure and Development of Solar Active Regions', *IAU Symp.* **35**, 267.
Smith, S. F. and Angle, K. L.: 1969, Final Rep. Lockheed, Palo Alto Res. Lab. LMSC 687340, June 1969.
Smolkov, G. Y.: 1966, *Studies in Geomagn. and Aeronomy*, 'Science', Publ. House, Moscow, p. 189.
Smolkov, G. Y.: 1970, *Res. Geomagn. Aeron. Solar Phys.* **15**, 49.
Sobolev, V. M.: 1958, *Bull. Astron. Obs. Pulkova* **20**, 12.
Sonnerup, B. U. Ø.: 1970, *J. Plasma Phys.* **4**, 161.
Spitzer, L.: 1956, *Physics of Fully Ionized Gases*, Interscience, New York.
Stellmacher, G.: 1969, *Astron. Astrophys.* **1**, 62.
Stepanov, V. E.: 1958, *Izv. Krymsk. Astrofiz. Observ.* **19**, 20.
Stepanov, V. E.: 1960, *Izv. Krymsk. Astrofiz. Observ.* **24**, 293
Stewart, R. T. and Sheridan, K. V.: 1970, *Solar Phys.* **12**, 229.
Stix, T. H.: 1962, *The Theory of Plasma Waves*. McGraw-Hill Book Co., New York, Ch. 4.
Stratton, J. A.: 1941, *Electromagnetic Theory*, McGraw-Hill Book Co., New York and London.
Sturrock, P. A. and Coppi, B.: 1964, *Nature* **204**, 61.
Sturrock, P. A. and Coppi, B.: 1965, *Astrophys. J.* **143**, 3.
Sturrock, P. A. and Woodbury, E. T.: 1967, in P. Sturrock (ed.), *Plasma Astro-Physics*, Academic Press, New York, p. 155.
Švestka, Z.: 1962, *Bull. Astron. Inst. Czech.* **13**, 190.
Swann, W. F. G.: 1933, *Phys. Rev.* **43**, 217.
Sweet, P. A.: 1958a, *IAU Symp.* **6**, 123.
Sweet, P. A.: 1958b, *Nuovo Cim. Suppl.* **8**, Ser. X, 188.
Sweet, P. A.: 1964, *AAS-NASA Symp. on the Physics of Solar Flares 1963*, p. 409.
Syrovatsky, S. I.: 1966, *Astron. Zh.* **43**, 340.
Syrovatsky, S. I.: 1969, in C. de Jager and Z. Švestka (eds.), 'Solar Flares and Space Research', *XI Cospar Symp.* North-Holland Publ. Co., Amsterdam, p. 346.
Syrovatsky, S. I.: 1970, in E. R. Dyer (general ed.), *Solar Terrestrial Physics 1970*, Part 1, D. Reidel Publ. Co., Dordrecht, p. 119.
Tandberg-Hanssen, E.: 1959, *Astrophys. J.* **130**, 202.
Tandberg-Hanssen, E.: 1960, *Astrophys. Norv.* **6**, 161.
Tandberg-Hanssen, E.: 1963, *Astrophys. J.* **137**, 26.
Tandberg-Hanssen, E.: 1964, *Astrophys. Norv.* **9**, 13.
Tandberg-Hanssen, E.: 1970, *Solar Phys.* **15**, 359.
Tandberg-Hanssen, E. and Anzer, U.: 1970, *Solar Phys.* **15**, 158.
Tandberg-Hanssen, E. and Zirin, H.: 1959, *Astrophys. J.* **129**, 408.
ten Bruggencate, P.: 1953, *Accad. Naz. Lincei, Convegno Volta* **11**, 163.
ten Bruggencate, P. and Elste, G.: 1958, *Nature* **182**, 1154.

Teske, R. G.: 1971, *Solar Phys.* **17**, 76.
Thomas, R. N.: 1948a, *Astrophys. J.* **108**, 130.
Thomas, R. N.: 1948b, *Astrophys. J.* **108**, 142.
Thomas, R. N.: 1957, *Astrophys. J.* **165**, 260.
Thomas, R. N. and Athay, R. G.: 1961, *Physics of the Solar Chromosphere,* Interscience Press, New York.
Tousey, R. and Koomen, M. J.: 1973, Proc. Conf. on Flare-Produced Shock Waves in the Corona and Interplanetary Space, NCAR, Boulder, Sept. 1972.
Uchida, Y.: 1963, *Publ. Astron. Soc. Japan* **15**, 65.
Uchida, Y.: 1968, *Solar Phys.* **4**, 30.
Uchida, Y.: 1970, *Publ. Astron. Soc. Japan* **22**, 341.
Uchida, Y., Altschuler, M., and Newkirk, G. N.: 1973, *Solar Phys.* **28**, 495.
Ulloa, A.: 1779, *Phil. Trans. Roy. Soc. (London)* **69**, 105.
Unno, W.: 1956. *Publ. Astron. Soc. Japan* **8**, 108.
Vaiana, G. S., Reidy, W. P., Zehnpfening, T., and Van Speybroeck, L.: 1968, *Science* **161**, 564.
Valniček, B.: 1964, *Bull. Astron. Inst. Czech.* **15**, 207.
Valniček, B.: 1968, in K. O. Kiepenheuer (ed.), 'Structure and Development of Solar Active Regions, *IAU Symp.* **35**, 282.
van de Hulst, H. C.: 1950, *Bull. Astron. Inst. Neth.* **11**, 150.
van de Hulst, H. C.: 1953, in G. K. Kuiper (ed.), *The Sun,* University of Chicago Press, p. 296.
Vassenius, B.: 1733, *Phil. Trans. Roy. Soc. (London)* **38**, 134.
Von Klüber, H.: 1961, *Monthly Notices Roy. Astron. Soc.* **123**, 61.
Vyazanitsyn, V. P.: 1947, *Izv. Glavnoi Astr. Obs.* **17**, No. 136, 1.
Vyssotsky, A. N.: 1949, *Meddelanden Lund Obs.* **2**, No. 126, 9.
Waldmeier, M.: 1937, *Z. Astrophys.* **14**, 91.
Waldmeier, M.: 1938, *Z. Astrophys.* **15**, 299.
Waldmeier, M.: 1941, *Ergebn. u. Probleme der Sonnen-Forsch.* Leipzig. p. 234.
Waldmeier, M.: 1946, *Experient.* **2**, 220.
Waldmeier, M.: 1949, *Z. Astrophys.* **26**, 305.
Waldmeier, M.: 1951, *Z. Astrophys.* **28**, 208.
Waldmeier, M.: 1952, *Z. Astrophys.* **30**, 150.
Waldmeier, M.: 1961, *Z. Astrophys.* **53**, 142.
Waldmeier, M.: 1970, *Solar Phys.* **15**, 167.
Waldmeier, M. and Müller, H.: 1950, *Z. Astrophys. J.* **27**, 58.
Wallenquist, A.: 1957, *Uppsala Astron. Obs. Ann.* **4**, 1.
Warwick, J. W.: 1955, *Astrophys. J.* **121**, 190.
Warwick, J. W.: 1957, *Astrophys. J.* **125**, 811.
Warwick, J. W.: 1962, *Publ. Astron. Soc. Pacific* **74**, 302.
Warwick, J. W.: 1965, in J. Aarons (ed.), *Solar System Radio Astronomy*, Plenum Press, New York, p. 131.
Warwick, J. W. and Hyder, C. L.: 1965, *Astrophys. J.* **141**, 1362.
Wentzel, D. G.: 1963, *Astrophys. J.* **137**, 135.
Wentzel, D. G. and Solinger, A. B.: 1967, *Astrophys. J.* **148**, 877.
Wesley, W. H.: 1927, *Mem. Roy. Astron. Soc.* **64**, Appendix.
Westin, H. and Liszka, L.: 1970, *Solar Phys.* **11**, 409.
Weyman, R.: 1960, *Astrophys. J.* **132**, 380.
Wild, J. P.: 1969, *Proc. Astron. Soc. Australia* **1**, 181.
Wild, J. P.: 1970, *Proc. Astron. Soc. Australia* **1**, 365.
Wild, J. P. and Zirin, H.: 1956, *Australian J. Phys.* **9**, 315.
Wild, J. P., Sheridan, K. V., and Kai, K.: 1968, *Nature* **218**, 536.
Williamson, N. K., Fullerton, C. M., and Billings, D. E.: 1961, *Astrophys. J.* **133**, 973.
Wurm, K.: 1948, *Ann. Phys.* **6**, 3, 139.
Yakovkin, N. A. and Zel'dina, M. Yu.: 1963, *Astron. Zh.* **40**, 847.
Yakovkin, N. A. and Zel'dina, M. Yu.: 1964a, *Astron. Zh.* **41**, 336.
Yakovkin, N. A. and Zel'dina, M. Yu.: 1964b, *Astron. Zh.* **41**, 914.
Yakovkin, N. A. and Zel'dina, M. Yu.: 1969, *Soln. Dann. Bull.* No. 4, 82.
Yakovkin, N. A. and Zel'dina, M. Yu.: 1971, *Solar Phys.* **19**, 414.
Yeh, T. and Axford, W. I.: 1970, *J. Plasma Phys.* **4**, 207.
Young, C. A.: 1896, *The Sun,* D. Appleton, New York.
Zanstra, H.: 1950, *Proc. Koninkl. Ned. Akad. Wetenschap* **53**, 1289.
Zanstra, H.: 1955, in A. Beer (ed.), *Vistas in Astronomy* **1**, 256.

Zel'dovich, Ya. B. and Raizer, Yu. P.: 1967, *Physics of Shockwaves and High-Temperature Hydro-dynamic Phenomena*, Vol. II, Ch. X, Academic Press, New York.
Zirin, H.: 1959, *Astrophys. J.* **129**, 414.
Zirin, H.: 1961, *Astron. Zhurn.* **38**, 861. (*Soviet Astron.* **5**, 660, 1962.)
Zirin, H.: 1964, *Astrophys. J.* **140**, 1216.
Zirin, H.: 1966, *The Solar Atmosphere*, Blaisdell-Ginn, Waltham, Mass.
Zirin, H.: 1968, *Nobel Symp.* **9**, 131.
Zirin, H.: 1972, *Solar Phys.* **22**, 34.
Zirin, H. and Acton, L. W.: 1967, *Astrophys. J.* **148**, 501.
Zirin, H. and Severny, A. B.: 1961, *Observatory* **81**, 155.
Zirin, H. and Tandberg-Hanssen, E.: 1960, *Astrophys. J.* **131**, 717.
Zirin, H. and Werner, S.: 1967, *Solar Phys.* **1**, 66.
Zirin, H., Hall, L. A., and Hinteregger, H. E.: 1963, in W. Priester (ed.), *Space Res.* **3**, North-Holland Publ. Co., Amsterdam, p. 760.
Zirker, J. B.: 1959, *Astrophys. J.* **129**, 424.
Zuikov, V. N.: 1955, *Izv. Glavnoi Astr. Obs.* **20**, No. 155, 22.

INDEX OF NAMES

INDEX OF SUBJECTS

GEOPHYSICS AND ASTROPHYSICS MONOGRAPHS

AN INTERNATIONAL SERIES OF FUNDAMENTAL TEXTBOOKS